THE BASICS OF ORGANIC CHEMISTRY

THE BASICS OF ORGANIC CHEMISTRY

MARTIN CLOWES

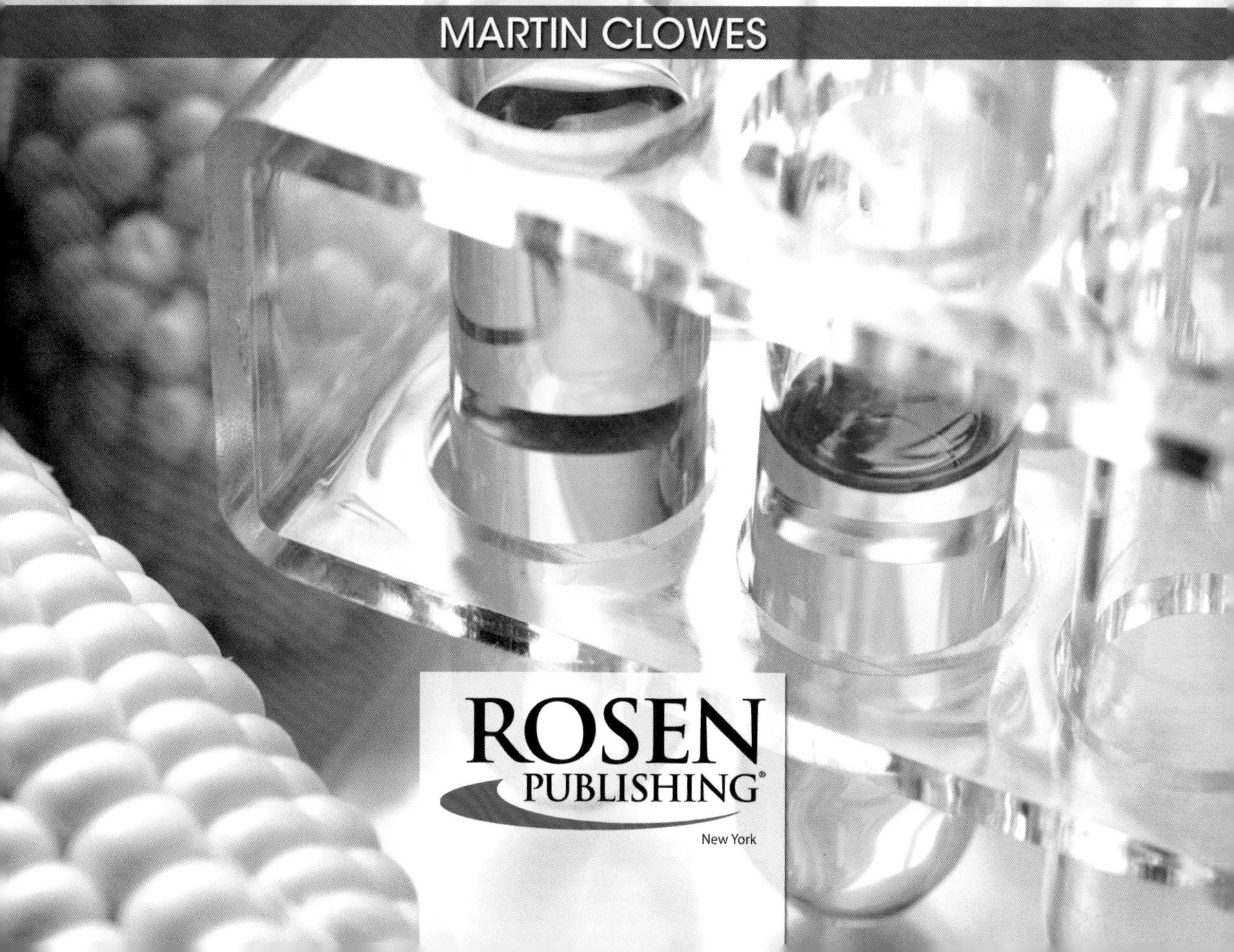

This edition published in 2014 by:

The Rosen Publishing Group, Inc.
29 East 21st Street
New York, NY 10010

Library of Congress Cataloging-in-Publication Data

Clowes, Martin, author.
The basics of organic chemistry/Martin Clowes. — First edition.
pages cm. — (Core concepts)
Audience: Grades 7 to 12.
Includes bibliographical references and index.
ISBN 978-1-4777-2714-0 (library binding)
1. Chemistry, Organic—Juvenile literature. I. Title.
QD251.3.C56 2014
547—dc23

2013026819

Manufactured in the United States of America

CPSIA Compliance Information: Batch #W14YA: For further information, contact Rosen Publishing, New York, New York, at 1-800-237-9932.

CONTENTS

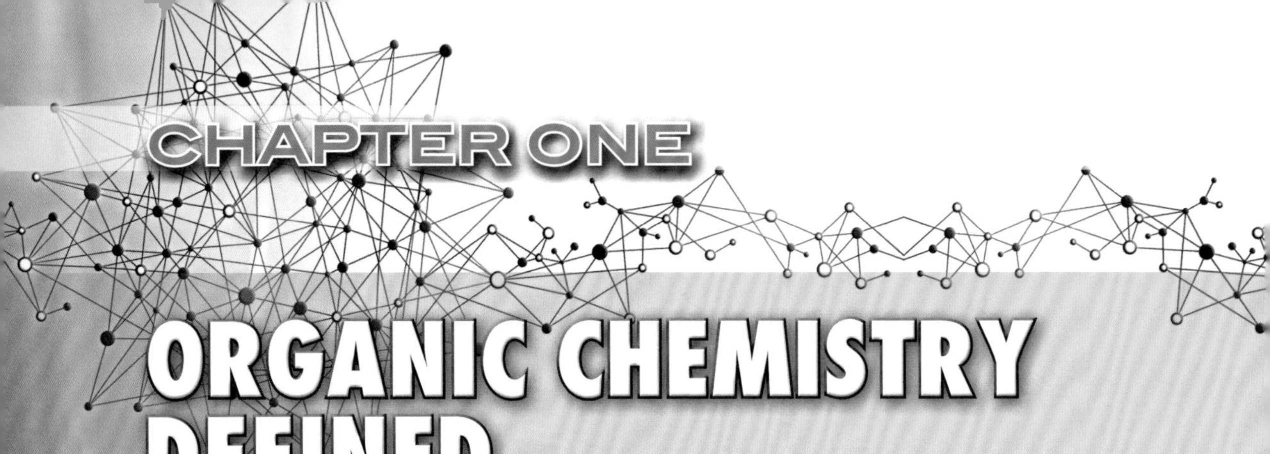

CHAPTER ONE

ORGANIC CHEMISTRY DEFINED

A colorful pattern is made when light shines on a layer of oil floating on water. Oil is an organic substance.

Chemists divide chemicals into two groups: organic and inorganic. Organic chemicals contain large amounts of carbon. They occur in everything from plastics to gasoline to drugs and even make up life-forms, including you!

When you are learning about chemistry, you are shown many examples of how atoms and molecules (atoms joined together) react with each other, and how their structures affect the way they behave. Most of these examples are very simple, so you can understand them easily. You learn about water, salt, and metals—substances that you come

KEY DEFINITIONS

- **Atom:** The smallest piece of an element that still retains the properties of that element.
- **Compound:** A substance formed when atoms of two or more different elements bond together.
- **Molecule:** Two or more atoms connected by chemical bonds.

HISTORY

DIVIDING CHEMISTRY

Chemists first studied organic compounds in the early 19th century, when people began investigating the substances inside the bodies of life-forms. Many believed that these compounds were so complex that they could only be made inside a living body, or organism. Because of this, Swedish chemist Jöns Jacob Berzelius (1779–1848) called the compounds organic. All other compounds were therefore inorganic. However, in 1828, the German chemist Friedrich Wöhler (1800–1882) showed that organic compounds could be made in a laboratory as well. He reacted two inorganic compounds together and, completely by accident, produced urea, a substance that occurs in urine. This discovery showed that organic compounds were built the same way as other compounds, but that they were just more complicated.

Jöns Jacob Berzelius is considered the first organic chemist, and he helped to determine the accurate atomic weights of many elements. Berzelius also came up with the system of chemical symbols that is used today.

across every day. However, most of the substances around you now are not so simple to make or easy to understand.

COMPOUNDS

Living bodies—the most complex things in nature—and many of the most useful materials made by people, such as plastics, fuels, and drugs, are made from very

Coal is a rock that contains organic chemicals. Coal is used mainly as a fuel, but it is also a source of useful chemicals.

complex compounds. Chemists describe the compounds as being organic. That is because those compounds that occur in nature have all originally been produced by living things.

A compound is a substance that is made when the atoms of two or more elements bond together. Organic compounds contain many atoms, perhaps hundreds of thousands, bonded together in a very precise pattern. All organic compounds are based on the element carbon (C). The compounds also contain atoms of other elements, most often hydrogen (H), but oxygen (O), nitrogen (N), and chlorine (Cl) are also commonly involved.

STUDYING ORGANIC COMPOUNDS

The first chemists to investigate organic compounds could not figure out much about them. The methods used for studying inorganic compounds did not work very well with organic ones. Chemists knew that organic compounds contained carbon and hydrogen because when the compounds burned, they produced water vapor (H_2O) and carbon dioxide (CO_2). Burning, or combustion, occurs when a compound reacts with oxygen. Chemists can calculate the proportions of carbon and hydrogen atoms in an organic compound by measuring

CHEMISTRY IN ACTION

CHEMISTRY AND LIFE

Chemical reactions keep all life-forms alive. It is chemical reactions that extract energy from food, cause muscles to move, build new body tissues as we grow, and repair the damaged parts. The chemistry of life involves organic compounds. These compounds are very complex to make, and understanding how they react is a science in itself, known as biochemistry.

Many of the organic compounds produced by living things will be familiar to you. They include sugars, fats, and proteins. Sugars belong to a group of compounds called carbohydrates. They are all made up of carbon, hydrogen, and oxygen. Fats are slightly more complex molecules with long chains of joined carbon atoms. Sugars and fats are used as fuel by life-forms.

Proteins are more complex. They are the building blocks of a body and are used, for example, to make muscles and skin. Proteins have very large molecules, which include nitrogen, sulfur, and phosphorus atoms, as well as carbon, hydrogen, and oxygen atoms. Perhaps the most important organic compound is DNA. This molecule forms long chains that carry genes—the coded plans for building a living body.

A pair of butterflies in the wild. Their bodies are made up of organic chemicals, such as sugars, proteins, and fats. All life on Earth is based on organic compounds.

the amount of each of these gases produced when it burns.

In 1828, Friedrich Wöhler discovered that organic compounds could be made from inorganic ones. Chemists began to look at organic compounds in a new way. They looked at simple compounds with just a few atoms in them. These included nut oils, formic acid made by stinging ants, and alcohol made by rotting fruit.

The chemists saw that some compounds react in the same ways even though they are very different in other ways. The scientists realized that these compounds all have the same group of atoms somewhere in their molecule. It is these so-called functional groups that give the compounds their properties. Today's organic chemists study how these functional groups work and even make up new ones.

KEY DEFINITIONS

- **Biochemistry:** The study of chemical reactions inside bodies.
- **Inorganic:** Describes a substance that is not organic.
- **Organic:** Describes a compound that is made of carbon and generally also contains hydrogen.

There is a wide range of everyday products that contains organic compounds, including nylon clothing.

CHAPTER TWO

CARBON ATOMS AND BONDING

All organic compounds contain carbon atoms. Carbon is the only element with atoms that can form limitless chains as well as branched and ring structures. This ability is a result of the way carbon forms bonds.

Organic compounds exist in a mind-boggling array of shapes and sizes. Their molecules often form chains, rings, and networks of the two, but there are also coiled molecules, spheres, and even tiny tubes. All this variety is a result of the ability of carbon atoms to form strong bonds. To understand how carbon atoms form so many molecules, it is worth looking at pure carbon itself.

FOUR FORMS OF PURE CARBON

Carbon occurs in nature in four main forms: soot, fullerenes, diamond, and graphite. Both soot and fullerenes are made when carbon-containing

Diamonds are made from pure carbon. The carbon atoms are connected in a rigid network, which makes diamond the hardest substance known.

KEY DEFINITIONS

- **Amorphous:** Something that has no definite shape.
- **Atom:** The smallest piece of an element that still retains the properties of that element.
- **Element**: A substance made up of just one type of atom.
- **Molecule:** Two or more atoms connected together.
- **Organic:** Describes a compound that is made of carbon and that generally also contains hydrogen.

compounds are burned. Fullerenes are very fragile structures and were only discovered 20 years ago.

Soot, a fine black powder also known as amorphous carbon, does not have an ordered structure; its carbon atoms are arranged randomly.

Graphite and diamond are the two most stable and familiar forms of pure carbon. Despite being made of nothing but carbon atoms, the two substances are very different.

Graphite is a black and shiny material. It glistens in the light slightly, like a metal. Also like a metal, graphite can conduct (carry) electric currents. Graphite is used in pencil leads because it is a very soft substance. When a pencil is moved across a piece of paper, the graphite wears away, leaving a dark line. This dark line is a very thin layer of carbon atoms.

In many ways, diamond is the opposite of graphite. It is see-through and colorless and cannot conduct electricity. Diamond is also extremely hard, the hardest of all substances, in fact. It is very hard to break a diamond. When diamond crystals do break, they split along flat surfaces. This makes it possible to make diamonds into attractive jewels. The points between the surfaces of a diamond are very sharp and hard enough to

A drop of crude petroleum, otherwise known as oil. Oil is a mixture of many carbon-containing compounds, such as tar and gasoline.

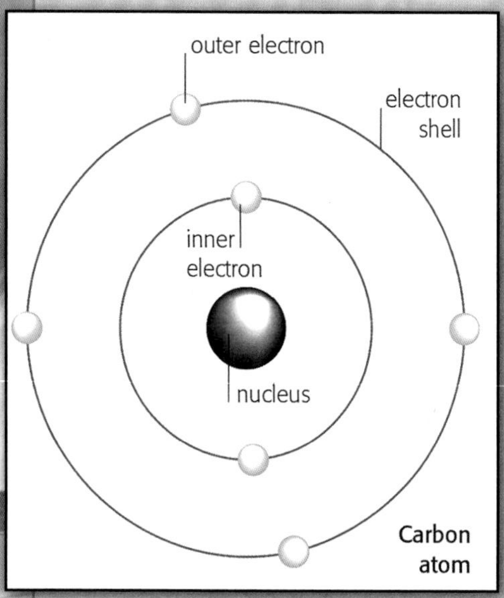

CHEMISTRY IN ACTION
THE CARBON ATOM

Atoms are made up of three types of particles: protons, neutrons, and electrons. Protons and neutrons form the central nucleus. Protons have a positive electric charge; neutrons have no charge. A carbon atom has six protons in its nucleus. Most carbon atoms also have six neutrons, although some rare types have seven or eight.

Particles called electrons move around the nucleus in two layers, or shells. Electrons are negatively charged and they are attracted to the positive charge of the protons in the nucleus. This force holds the atom together. Carbon atoms have six electrons—the same number as protons. Two electrons move in the inner shell close to the nucleus. The other four reside in the outer shell. An atom's outer electrons are involved in chemical bonds. Carbon atoms can form four bonds with other atoms. This ability is the key to carbon's remarkable chemistry.

cut through any solid. Many drills and saws have diamond tips.

ATOMIC STRUCTURE OF CARBON

How can the atoms of one element make two so very different materials? The answer is in the way the atoms are connected inside each substance. To understand how carbon forms bonds, we must look inside an atom of carbon.

Carbon atoms have four electrons in their outer shell. These electrons are the ones that form bonds with other atoms. Atoms form bonds by sharing, taking, or giving away their outer electrons. They do this to make their outer electron shell full, which makes the atoms stable.

An atom's outer shell can hold eight electrons. To become stable, a carbon

Bubbles of carbon dioxide (CO_2) gas fizz out of a can as the soda is poured into a glass. Carbon dioxide and many other carbon compounds are classed as inorganic. However, the carbon atoms in the compounds form bonds in the same way as they do in organic compounds.

A CLOSER LOOK
CARBON AND COVALENT BONDS

A carbon atom can form up to four covalent bonds. These bonds involve two atoms sharing electrons. In a simple covalent bond, each atom provides one electron, forming a pair. The pair of electrons sits in the outer shell of both atoms. As a result, the atoms are pulled side by side. The shared pair of electrons is being pulled on by the positive charge of the nucleus of both atoms. These pulling forces hold, or bond, the atoms together. This arrangement is called a single bond.

A carbon atom can form two or three bonds with one other atom. These are known as double and triple bonds. Most of the time, double and triple bonds form between two carbon atoms.

In a double bond, each atom shares two of its electrons. A triple bond involves three pairs. Compounds with double and triple bonds are more reactive than those with single bonds. The bonds often break so that they can form extra more stable single bonds.

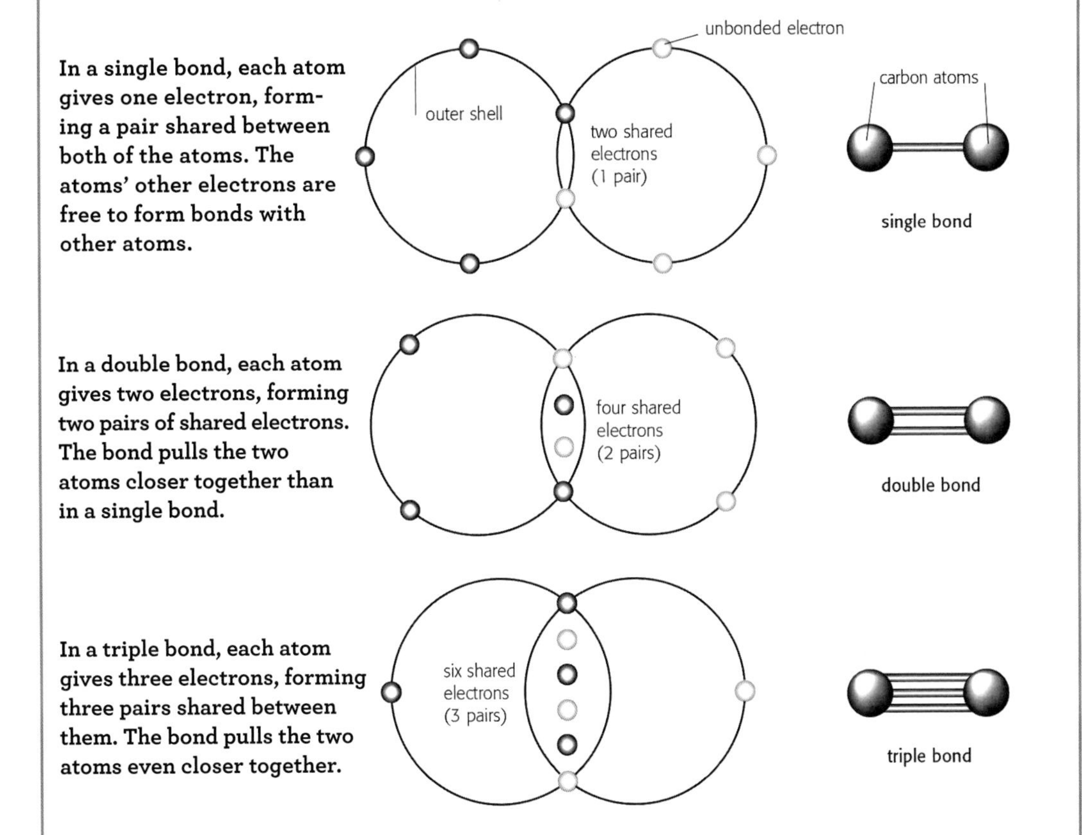

In a single bond, each atom gives one electron, forming a pair shared between both of the atoms. The atoms' other electrons are free to form bonds with other atoms.

In a double bond, each atom gives two electrons, forming two pairs of shared electrons. The bond pulls the two atoms closer together than in a single bond.

In a triple bond, each atom gives three electrons, forming three pairs shared between them. The bond pulls the two atoms even closer together.

atom must share four electrons with other atoms. A bond formed when atoms share electrons is called a covalent bond. Carbon atoms are unusual, however, because their outer shell is half full (or half empty). That makes the atoms more stable than most. As a result, carbon atoms can form two or even three strong bonds with just one other atom.

Carbon's ability to form so-called double and triple bonds is behind the differences between graphite, diamond, and other forms of pure carbon.

KEY DEFINITIONS

- **Covalent bond:** A bond in which two or more atoms share electrons.
- **Electron shell:** A layer of electrons that surrounds the nucleus of an atom.
- **Inorganic:** Describes a substance that is not organic.
- **Nucleus:** The central core of an atom containing protons and neutrons.

TYPES OF BONDS

Inside a diamond, carbon atoms are connected by only single bonds. Each carbon atom is bonded to the four atoms surrounding it. With all the atoms bonded to one another, a piece of diamond is one huge molecule.

Diamonds are measured in a unit called carats. A one-carat diamond weighs 0.007 ounces (0.2 g). A diamond this size has 10^{23} atoms (the number 10 followed by 23 zeros). In reality, no diamond is perfect; there are always some tiny cracks in the crystal.

Diamond's extreme hardness is a result of its atoms being bonded into a rigid interconnecting structure. Graphite is so soft and different from diamond in many other ways because some of the atoms inside are joined by a weaker type of bond.

Inside graphite, each carbon is joined to just three atoms by single bonds. The atom is also connected to a fourth atom, but this time the bond is a weak bond that forms in the same way as a double

Thick smoke flows out of a large smokestack. Smoke is hot gases that have fine particles of solids mixed into them. Most of the solids are particles of soot. Unlike other forms of carbon, the atoms inside soot are not organized in any ordered patterns.

Graphite is most familiar as the substance in pencil leads.

A CLOSER LOOK

TYPES OF PURE CARBON

Pure carbon occurs in more than one form. Each form is called an allotrope. The three carbon allotropes are graphite, diamond, and fullerenes. Graphite molecules form as sheets. Each carbon atom is bonded to three others, and they form interconnecting hexagons (six-sided shapes). Each atom also forms a fourth, weaker bond with an atom in a neighboring sheet. As a result, the sheets can move past each other easily, making graphite soft. In diamond, all the bonds are equal in strength. That makes it very difficult to break a diamond apart because there are no weak points in the structure. The last form, fullerenes, are very fragile. They can be described as a single sheet of graphite rolled into a ball.

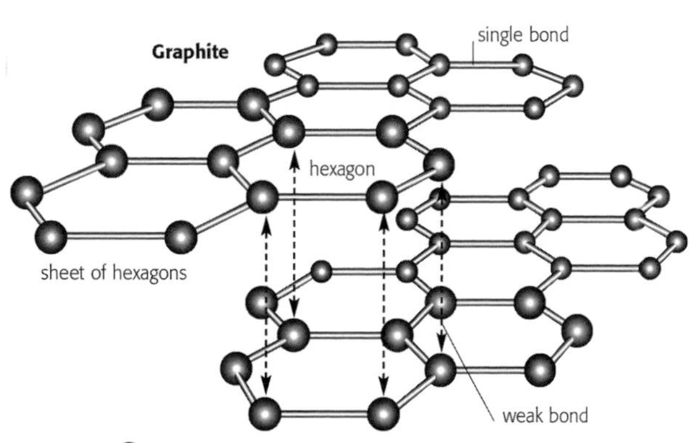

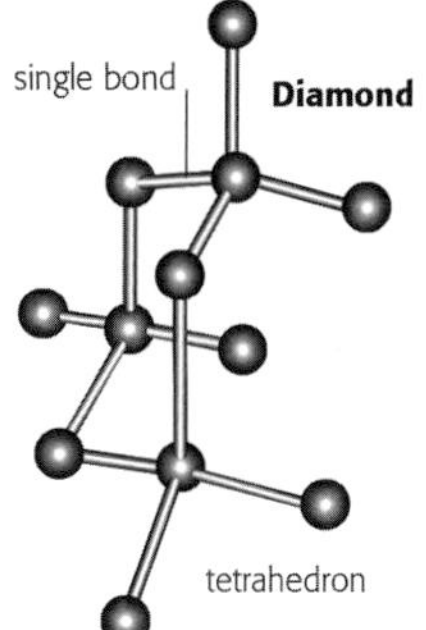

▲ In graphite the carbon atoms form sheets of hexagons. Each atom is joined to the sheet above or below by a weak bond. ◀ In diamond, each set of four carbon atoms forms a pyramid-like shape called a tetrahedron. The pyramids are all connected to each another. ▼ This fullerene is a sphere of 60 carbon atoms.

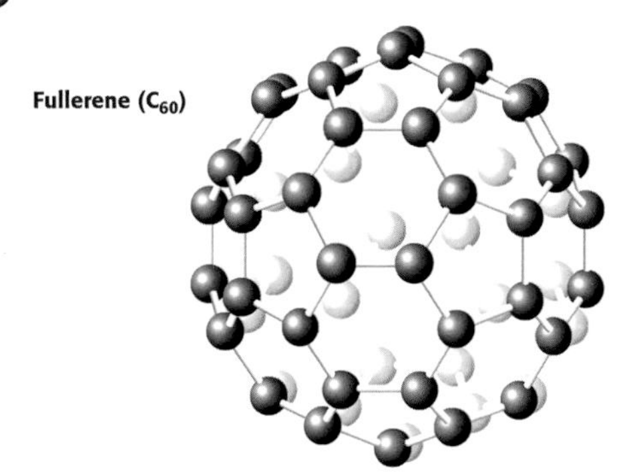

Diamonds are used as jewels because they reflect light in such a way that they sparkle.

A CLOSER LOOK AT OTHER ALLOTROPES

Carbon is not the only element to have allotropes—more than one pure form. Pure oxygen, sulfur, tin, and arsenic are some examples of other elements that exist in several forms. Although an element's allotropes look different and have different properties, they are all made up of just one type of atom. However, the arrangement of atoms is different inside each allotrope, and that is what gives each form its different properties.

For example, arsenic exists as gray arsenic and yellow arsenic. The first form is hard and shiny like a piece of metal. However, yellow arsenic is a crumbly powder. Sulfur has several allotropes. Sulfur crystals form many shapes, including spikes and cubes.

Tin is a metal with two allotropes. When it warms up, the element becomes white tin. As it cools down again, it slowly turns back into gray tin. Perhaps the most familiar allotrope is ozone. This is a form of oxygen.

bond. However, the bond in graphite is not quite the same because only one pair of electrons is shared.

This fourth bond is very weak, and as a result the carbon atoms inside graphite are not as strongly connected to each other. When a force pushes on graphite, it breaks the atoms' weak bonds easily, and the graphite breaks or changes shape. A lump of graphite feels slippery. That is because even the touch of your fingers is enough to rub away a layer of graphite. Graphite is used as a lubricant instead of oil or grease.

FREE ELECTRICITY

The structure of graphite also explains how it carries electricity, though diamond cannot. An electric current is a flow of electrons—sometimes other charged particles—through a substance. The moving particles transfer energy from one place to another, and electric currents are used

Not all carbon-containing compounds are organic. Some simple carbon compounds are classed as inorganic because their carbon atoms do not form chains or rings. These cliffs are made from limestone, which mostly contains calcium carbonate ($CaCO_3$). In this compound, each carbon atom is bonded to three oxygen atoms, and together they are joined to a calcium atom. Calcium carbonate is a very useful substance. People use it to make chalks, steel, and in construction.

to power many machines in our homes, schools, and places of work.

Substances that can carry electricity are conductors; they have electrons that are free to move around inside. Insulators—materials that do not carry electricity—do not have free electrons. Graphite is a conductor because the electrons involved in the weak bonds break free easily. They then flow through the graphite crystal between the sheets of carbon atoms. All the electrons in diamond are held in strong bonds and cannot be released to form a current. As a result, diamond is an insulator.

A THIRD CARBON ALLOTROPE

Fullerenes, the third structural form of carbon, are also conductors. However, the way their electrons are free to move is different again from other carbon allotropes. Looking at the structure of fullerenes will also help us understand the properties of organic compounds.

HISTORY

BALLS FROM FIRE

Diamond and graphite have been known about for thousands of years. However, a third form of carbon was discovered only 20 years ago. In 1985, English scientist Harold Kroto (1939–) teamed up with two U.S. chemists Richard Smalley (1943–) and Robert Curl (1933–). The trio were trying to figure out what the surface of a star might be like. They used a superhot laser to burn samples of carbon and then analyzed what was produced.

Their experiments produced a lot of clusters of carbon atoms, like the ones seen in soot. However, to their surprise they found that clusters containing 60 carbon atoms were also produced. These clusters were much bigger than they expected and did not break apart easily. The scientists realized that the carbon atoms must be forming a hollow, cagelike ball. Further experiments showed that balls and other hollow structures could be made with larger numbers of carbon atoms.

Kroto, Smalley, and Curl had discovered a new carbon allotrope and they won the Nobel Prize in Chemistry in 1996. They named the substances fullerenes after the U.S. architect R. Buckminster Fuller (1895–1983). In the 1950s, Fuller designed domes that, by chance, had the same shape and structure as the chemicals.

A geodesic dome built by Buckminster Fuller in Montreal, Canada, in 1967.

Fullerenes are made when carbon compounds are burned. These molecules are fragile, and in normal conditions they soon fall apart and form sootlike substances.

The smallest and simplest fullerene contains 60 carbon atoms. Its formula is C_{60}. This fullerene was the first to be discovered in 1985. It was named buckminsterfullerene for the designer of geodesic domes, which it resembles. All similar carbon structures are now referred to as fullerenes, and C_{60} has been nicknamed "the buckyball."

In a buckyball and other fullerenes, each carbon atom is bonded to three others. Most form into hexagons in the same way as in sheets of graphite. However, in a few cases the carbons also form pentagons (five-sided shapes). This sheet of interconnected hexagons and pentagons curves into a sphere.

Unlike in diamond and graphite, the carbon atoms in fullerenes do not form a fourth bond. Instead the spare, unbonded electrons from each atom are shared between them all. This creates a "cloud" of electrons that spreads evenly over the surface of the ball. The electrons in this cloud are free to move and carry an electric current.

It is hoped that fullerenes will be very useful substances. They have been made into nanotubes. Perhaps one day

KEY DEFINITIONS

- **Allotrope:** One form of a pure element.
- **Conductor:** A substance that carries electricity and heat well.
- **Crystal:** A solid made of repeating patterns of atoms.
- **Insulator:** A substance that does not transfer an electric current or heat.

CHEMISTRY IN ACTION

NANOTUBES

Fullerenes do not have to be balls. In 1991, Japanese scientist Iijima Sumio (1939–) made fullerenes that were tube shaped. The tubes were made from a sheet of carbon atoms bonded in the same hexagon pattern as graphite molecules. The structures were named nanotubes. Nanotubes are very thin. One long enough to stretch from Earth to the moon could be rolled into a ball the size of a poppy seed! So far, scientists can only make short pieces. If we learn how to make them long enough, there will be many uses for nanotubes. For example, the tubes could be woven to make a material many times stronger than steel but much lighter.

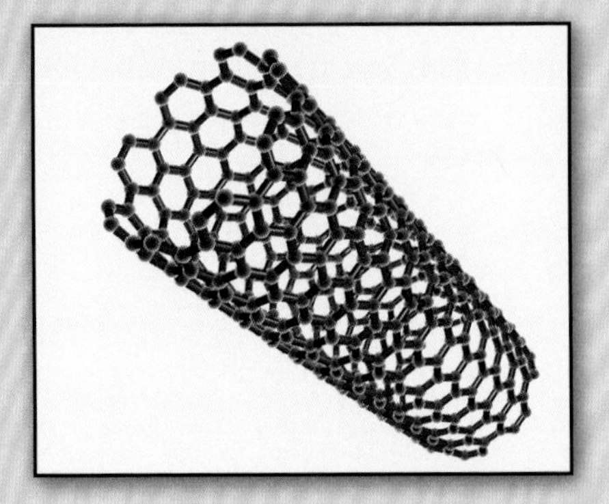

An illustration of a section of nanotube.

TAKING A CLOSER LOOK

COUNTING ATOMS

Atoms are too small to count one by one. How do chemists know how many atoms are contained in something? Chemists count atoms in moles. A mole is a very large number. Just as one dozen means 12, a mole means 602,213,670,000,000,000,000,000. Chemists weigh a substance to calculate how many moles of atoms or molecules it contains.

The atoms of each element have a fixed number of particles in their nucleus. Therefore, the atoms also have a fixed mass. (Electrons are so tiny, they do not really affect the atom's mass.) Carbon atoms have 12 times as many particles in their nucleus as hydrogen atoms. Therefore a mole of carbon atoms weighs 12 times as much as a mole of hydrogen atoms. The masses of all atoms can be compared in the same way. Chemists have agreed that carbon will be the benchmark for measuring moles. One mole of carbon weighs 12 grams (0.42 ounces). The mass of a mole of all other elements is based on this measurement.

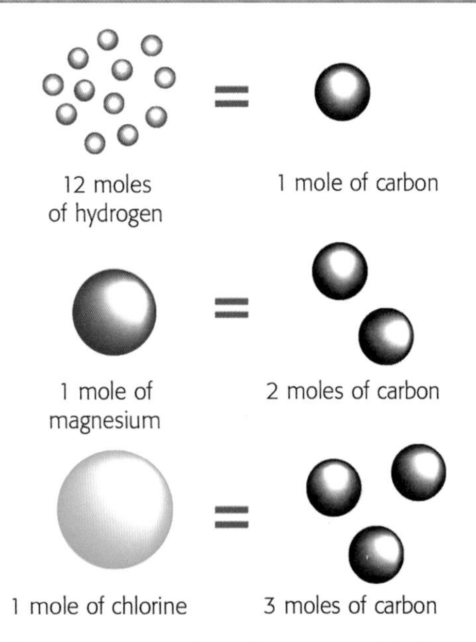

A diagram showing how the masses of different elements compare with the mass of carbon.

fullerenes will make pipes and wires in tiny machines. Fullerenes are hollow and can hold other atoms inside them. The atom inside is not bonded to the fullerene, so the two do not form a compound. Chemists have had to come up with a new way of describing such arrangements. A helium atom (He) inside a buckyball is written as $He@C_{60}$.

Plastic objects are made from long, chainlike organic compounds that can be molded into any shape.

CARBON-BASED COMPOUNDS

The ability of carbon atoms to form different types of bonds is what makes it possible for so many different organic compounds to exist. As you will see, organic compounds include everything from exploding gases to smelly oils to bouncy solids.

CHAPTER THREE

HYDROCARBONS AND CARBON CHAINS

Candle flames are produced when wax melts and burns in the air. The wax is made from a mixture of hydrocarbons called paraffins.

It is the ability of carbon atoms to form into chains that makes organic compounds so varied. There is no limit to the length of these chains, and they might branch into highly complex networks.

The simplest organic compounds contain just carbon (C) and hydrogen (H) atoms. These compounds are called hydrocarbons. Hydrocarbons are very useful substances. They occur mixed together in petroleum oil and natural gas. Gasoline and other fuels are examples of hydrocarbons. Hydrocarbons are also used to make thousands of other products.

Inside a hydrocarbon molecule, a carbon atom can be bonded to other carbon atoms or it can be connected to hydrogen atoms. Each carbon atom can bond to up to four other atoms. A hydrogen atom, on the other hand, can only form one bond. The hydrogen atoms in hydrocarbon molecules are always bonded to carbon atoms.

VERY STABLE BONDS

Hydrocarbons are covalent compounds. The atoms in their molecules are bonded because they are sharing electrons.

Carbon atoms can form into long chains because the bond between two carbon atoms is very stable. That is because carbon atoms have an outer electron shell that is half full.

The bond between a hydrogen and a carbon atom is also a strong one. Hydrogen atoms form stable bonds for the same reason that carbon atoms do. Hydrogen atoms have one electron shell. This shell can hold only two electrons, and hydrogen atoms have one. As a result, their electron shell is half full, just like in a carbon atom.

Hydrogen atoms can only form one bond each, so they cannot form chains. However, hydrogen forms the most complex and varied set of compounds in chemistry, when bonded with carbon.

KEY DEFINITIONS

- **Compound:** A substance formed when atoms of two or more different elements bond together.
- **Covalent bond:** A bond that forms between two or more atoms when they share electrons.
- **Hydrocarbon:** A type of organic compound containing only carbon and hydrogen atoms.
- **Molecule:** Two or more atoms connected together

SINGLE BONDS: ALKANES

The simplest hydrocarbon compounds are made of atoms connected by single bonds. This group of hydrocarbons is

Traffic runs along busy roads during an evening rush hour. Most automobiles and trucks are fueled with hydrocarbons, such as gasoline and diesel oil.

called the alkanes. Being made up of single bonds, alkane molecules have the same pyramid structure of diamond. However, the atoms form chains instead of a rigid network as in diamond. The shapes of alkanes and other organic compounds are complex, so all molecules are shown here as flat diagrams.

The smallest alkane is methane. This has the formula CH_4. A carbon atom is bonded to four hydrogen atoms. The next alkane is called ethane (C_2H_6). It has two carbon atoms bonded together. Each carbon atom is attached to three hydrogen atoms. The next compound, propane (C_3H_8), has three carbons in a row, while butane (C_4H_{10}) has four.

Alkane molecules get larger by adding more carbon atoms. The compounds have the general formula $C_nH_{(2n+2)}$, where n is the number of carbon atoms in a molecule. For example, in methane n equals 1, so the number of hydrogen atoms is $(2 \times 1) + 2 = 4$.

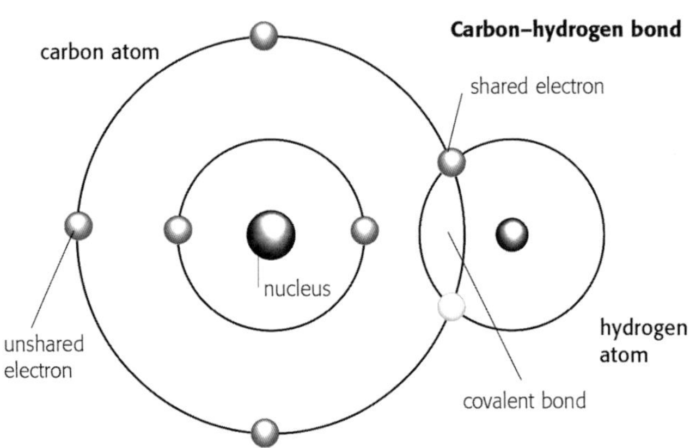

The covalent bond between a carbon and hydrogen atom.

NAMES OF MOLECULES

Number of carbon atoms	Prefix
1	Meth-
2	Eth-
3	Prop-
4	But-
5	Pent-
6	Hex-
7	Hept-
8	Oct-
9	Non-
10	Dec-

NAMING ORGANIC COMPOUNDS

With so many compounds to understand, chemists have developed a system for naming organic compounds. They have agreed on a prefix (the beginning of a word) for each number of carbon atoms in a molecule. For example, the prefix for two atoms is eth-, while molecules with eight atoms begin with oct-. All alkane compounds end in -ane. Therefore, C_2H_6 is called ethane and C_8H_{18} is named octane. The same system is used for other groups of hydrocarbons.

REACTIONS OF ALKANES

As we have learned, the bonds in alkane molecules are very stable. As a result, alkane compounds are not very reactive. Their most important reaction is

Hydrocarbons are not just found on Earth. A recent probe to Titan, a moon of the planet Saturn, found clouds and pools of methane on the moon. This is the probe's view as imagined by an artist.

combustion, or burning. Combustion reactions occur when compounds react with oxygen.

When alkanes burn, they release a lot of energy. That is why they make good fuels. For example, the natural gas extracted from underground is mainly methane. This gas is used in cooktops, ovens, and boilers, and is also burned in power plants to make electricity.

The combustion reaction also produces carbon dioxide (CO_2) and water (H_2O). All hydrocarbons produce these compounds when they burn, but different hydrocarbons produce them in

KEY DEFINITIONS

- **Alkane:** A hydrocarbon chain in which all atoms are connected by single bonds.
- **Chemical equation:** Symbols and numbers that show how reactants change into products during a reaction.
- **Combustion:** The reaction that causes burning. Combustion is generally a reaction with oxygen in the air.

TAKING A CLOSER LOOK
THE ALKANES

Alkane molecules are made using only single bonds. All carbon atoms in the molecules are bonded to the maximum of four other atoms. Many of the most familiar hydrocarbons are alkanes. Gasoline fuel contains a lot of octane (C_8H_{18}). Paraffin wax used to make candles is a mixture of alkane compounds, each containing between 22 and 27 carbon atoms.

▼ *The three simplest alkanes.*

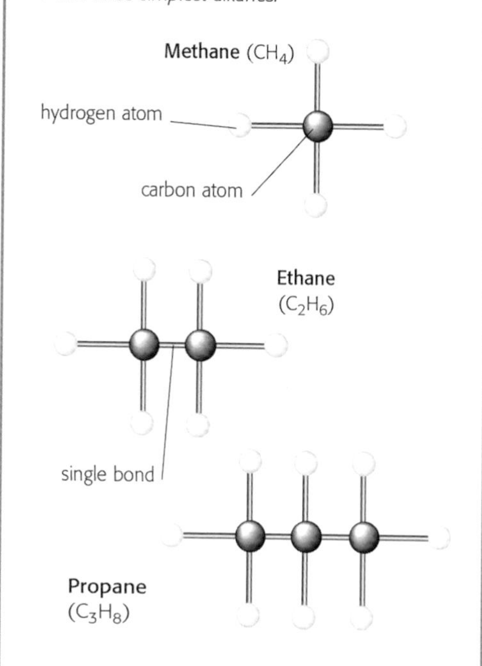

CHEMISTRY IN ACTION

HEATING EARTH

Gasoline and other fuels are often called "fossil fuels." That is because the petroleum oil, natural gas, and coal located deep beneath the ground are the remains of plants and other life-forms that died millions of years ago. Over the years, their remains have been buried, then heated and squeezed under the ground until they broke down into hydrocarbons. The carbon these compounds contain was taken from the air when the life-forms were alive. They used it to build their bodies.

For millions of years, that carbon has been locked underground. People today bring fossil fuels to the surface and burn them as fuels. When the hydrocarbons burn, they react with oxygen and produce carbon dioxide and water. These products are released into the air.

Nobody knows what a warmer world will be like. Some places will dry out into deserts, and other areas will be flooded by the sea.

People have been burning fossil fuels for about 200 years. In that time, the amount of carbon dioxide in Earth's atmosphere has increased by 50 percent. Earth's atmosphere acts like a blanket around the planet. It traps heat and keeps the world warm. However, now that the atmosphere has more carbon dioxide, it seems to be trapping too much heat.

The world is warming up because of all the carbon dioxide we are releasing as we burn fuels. Experts think that in 100 years, Earth will be about 5 degrees Fahrenheit (5°F; 3°C) warmer. That is likely to cause huge changes to the world's weather.

Oil and fats, such as those in butter, belong to a group of hydrocarbons called lipids. Oils are unsaturated lipids, while fats are saturated lipids.

different amounts. The equation for the combustion of methane is:

$$CH_4 + 2O_2 \rightarrow CO_2 + 2H_2O$$

DOUBLE BONDS: ALKENES

In an alkane, each carbon atom has four single bonds. Chemists say that alkanes are saturated hydrocarbons: the atoms inside are all bonded to the maximum number of other atoms. A molecule that contains carbon atoms that are bonded to less than four other atoms is described as being unsaturated.

Hydrocarbons that contain carbon atoms bonded to just three other atoms are called alkenes. The simplest alkene is ethene (C_2H_4). In a molecule of ethene, the two carbon atoms are connected by a double bond. With two bonds used up connecting to the other atom, some of the carbon atoms have just two bonds left for hydrogen atoms.

The alkenes increase in size in the same way as the alkanes: propene (C_3H_6) contains three carbon atoms, butene (C_4H_8) contains four, and so on. As with

TAKING A CLOSER LOOK

THE ALKENES

A hydrocarbon containing carbon atoms that are joined by double bonds is called an alkene. The double bonds fix the shape of an alkene molecule. Single bonds allow sections of the molecule to spin around independently. The double bond cannot rotate, so sections of the molecule cannot move. This has the most effect on the structures of branched molecules, where branches are attached on certain sides of the double bond.

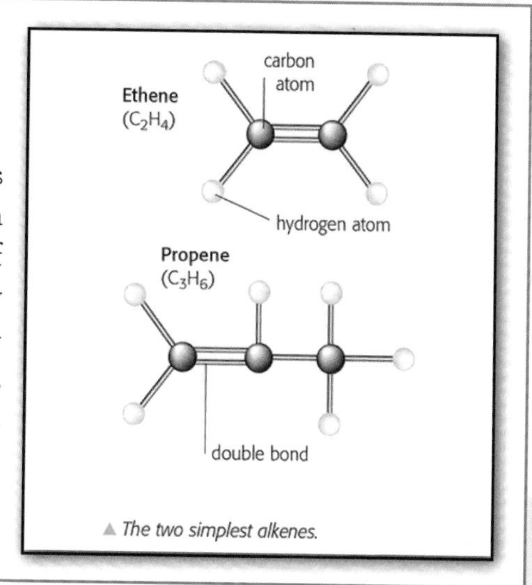

▲ *The two simplest alkenes.*

the alkanes, there is no limit to the length of an alkene chain.

REACTIONS OF ALKENES

Alkenes occur in petroleum mixed in with alkanes and other hydrocarbons. However, people also make alkenes because the double bonds make alkenes useful. Whereas alkanes are just burned as fuel, alkenes can be reacted with other compounds to make many products.

Alkenes are reactive because a double bond readily breaks to form two single bonds. For example, alkenes will react with hydrogen gas (H_2) to become alkanes. The reaction that turns ethene into ethane has the following equation:

$$C2H_4 + H_2 \rightarrow C_2H_6$$

This reaction is called an addition reaction because hydrogen has been added to the molecule.

TRIPLE BONDS: ALKYNES

Hydrocarbons that contain triple bonds between two carbon atoms are called alkynes. This group of compounds is more reactive than the alkenes. That is because a triple bond breaks to form three single bonds even more easily than a double bond does. Alkynes are so reactive, they are not very common in petroleum oil. Instead people make alkynes. The simplest alkyne is ethyne (C_2H_2), also called acetylene.

Like the alkenes, alkynes are used to make useful chemicals, such as plastics and medicines. However, ethyne is also used in torches for welding and cutting

TAKING A CLOSER LOOK

THE ALKYNES

Alkynes are hydrocarbons that have carbon atoms connected to each other by triple bonds. The simplest alkyne is ethyne. In this molecule, each carbon is bonded to just one hydrogen. Larger alkyne molecules do not have triple bonds between all the carbon atoms. Just one triple bond is enough to make them alkynes.

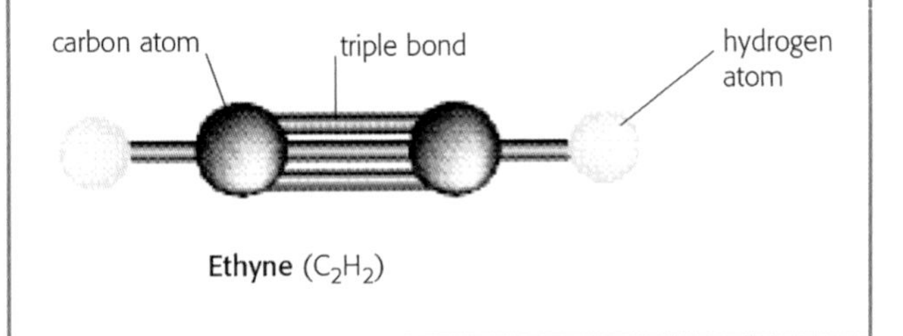

Ethyne (C_2H_2)

KEY DEFINITIONS

- **Alkene:** A hydrocarbon chain in which at least two carbon atoms are connected by a double bond.
- **Alkyne:** A hydrocarbon chain in which at least two carbon atoms are connected by a triple bond.
- **Saturated:** A hydrocarbon molecule in which all the carbon atoms are bonded to four other atoms. An unsaturated molecule contains double or triple bonds between carbon atoms.

metal. It burns with oxygen at up to 6,300 degrees Fahrenheit (6,300ºF; 3,500ºC)—much hotter than other fuels.

REFINED HYDROCARBONS

Before hydrocarbons such as alkanes and alkenes can be used for fuels or in industry, they must be refined. Refined hydrocarbons are called petrochemicals.

The main source of hydrocarbons is petroleum. This is a mixture of gases, liquids, and sludgelike solids. The word *petroleum* comes from the two Latin words for "rock" and "oil." Most petroleum is deep underground and it must be pumped to the surface. Petroleum is the remains of life-forms that have become buried under rocks over millions of years.

A refinery where petrochemicals are produced.

BRANCHED MOLECULES

Not all hydrocarbons are straight chains. Molecules with four or more carbon atoms can divide into branches. A branched molecule may contain the same number of atoms as a straight molecule, and their chemical formulas will be the same. Chemists use the naming system to describe how each molecule is organized instead.

A molecule's name is determined by the size of its longest straight chain. Alkane 1 (below) has four carbon atoms in a single chain. It is therefore named butane.

Alkane 2 also has four carbon atoms. However, its longest chain has three carbon atoms (like propane). The other carbon and its three hydrogen atoms ($-CH_3$, or methyl) are attached to the middle of the chain. The molecule is named methyl propane.

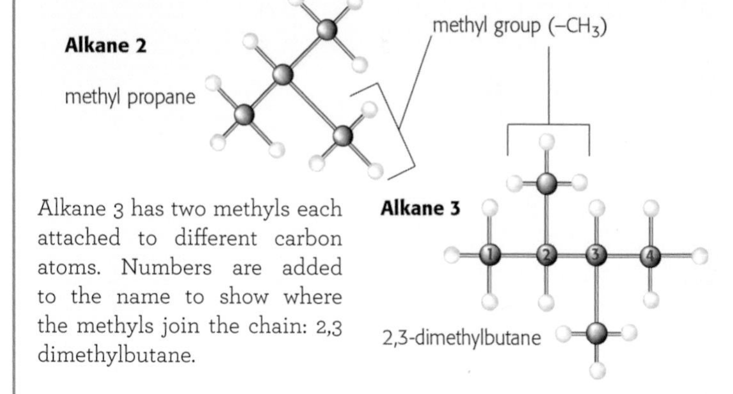

Alkane 3 has two methyls each attached to different carbon atoms. Numbers are added to the name to show where the methyls join the chain: 2,3 dimethylbutane.

Methyls and other branches are called alkyl groups. Their names are based on the number of carbon atoms they contain:

Number of carbons	Alkyl group	Formula
1	Methyl-	$-CH_3$
2	Ethyl-	$-C_2H_5$
3	Propyl-	$-C_3H_7$
4	Butyl-	$-C_4H_9$

OIL REFINING PROCESS

Unrefined petroleum is called crude oil. Once at a refinery, any gases, water, and unwanted solids, such as mud, are removed. The hydrocarbons that remain are then pumped into the bottom of a tall tower and heated to 720°F (380°C).

The tower is a fractional distillation column. It is used to separate the different sizes, or fractions, of hydrocarbon molecules. Heating the petroleum makes most of the hydrocarbons boil and turn into a gas.

The mixture of gases flows up the column. As the gas rises, it begins to cool down and turn into liquids. These liquids are collected at several points inside the column. Small and light hydrocarbon molecules, such as pentane (C_5H_{12}), have lower boiling points than large and heavy ones. The light molecules stay as gases until they get to the top of the column, where they are collected. Heavier fractions turn to liquid at points lower down, where they are collected.

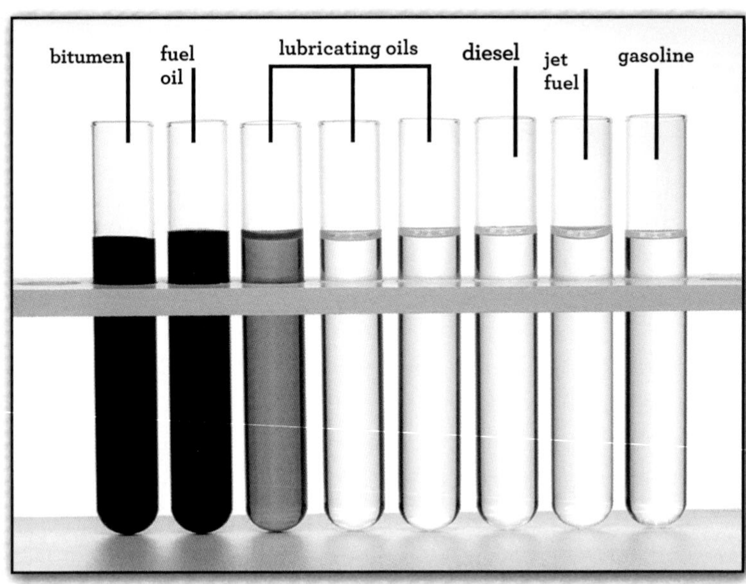

Fractions refined from crude oil. Bitumen has the highest boiling point. Gasoline has the lowest.

These are fractional distillation columns at a refinery.

CRACKING

Most of the hydrocarbons in crude oil are alkanes with straight chains. Around 90 percent of crude oil ends up as fuel—mainly gasoline for cars. However, after the fractions have been separated, only about 20 percent of them are useful immediately. The rest are pumped into chambers called reactors, where they are converted into more useful molecules.

Inside the reactors, the hydrocarbons are "cracked." Cracking is a process that breaks long alkane molecules into shorter alkanes and alkenes. Cracking requires the hydrocarbons to be very hot and under high pressure, but that alone is not enough. Cracking reactions require catalysts. A catalyst is a substance that helps a reaction along but is not changed after the reaction has finished.

A colorized image from a scanning electron microscope shows zeolite crystals that are used as a catalyst for making petrochemicals. This particular shape of crystal is used to "crack" long alkanes into smaller, branched molecules.

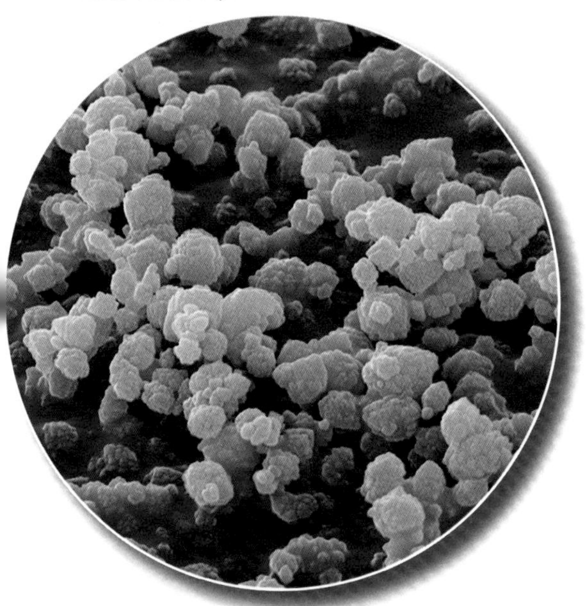

TAKING A CLOSER LOOK
A CATALYST AT WORK

Some chemical reactions occur only very slowly or require very high temperatures. A catalyst is a substance that makes a reaction run more quickly and so saves time and money. It works by bringing the reactants (the reaction's ingredients) together so that they can rearrange into the products. However, a catalyst is not used up by the reaction, so it can keep on working with new reactants.

Catalysts are also used to make sure that the correct reactions occur instead of other unwanted ones. Solid catalysts are used so that they do not get mixed up with gas or liquid reactants. For example, a solid cobalt catalyst is used to react methane with oxygen to make larger alkane molecules, such as ethane:

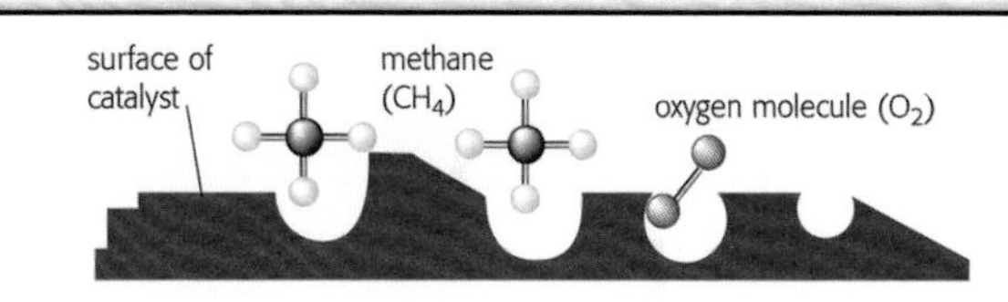

▲ *The molecules are brought together on the surface of the catalyst.*

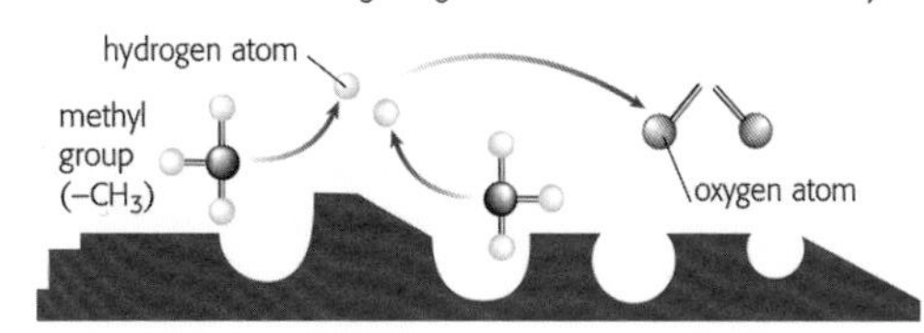

▲ *The oxygen molecule splits into single atoms. One oxygen atom pulls a hydrogen from each methane molecule, turning them into methyl groups.*

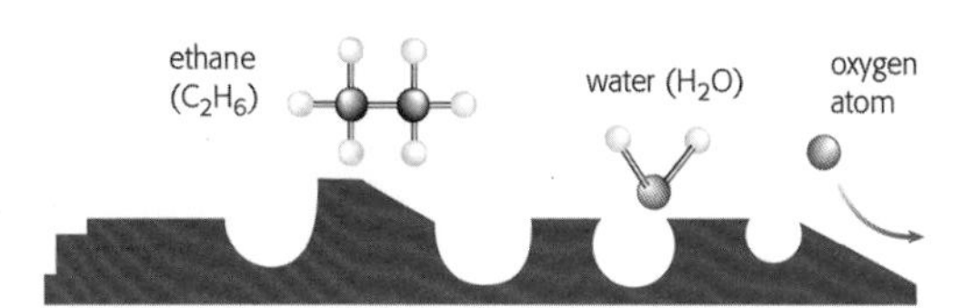

▲ *The two methyl groups bond together to form a molecule of ethane. The hydrogen atoms bond to the oxygen to make water. The second oxygen atom reacts with other methane molecules.*

The catalysts used in cracking are zeolites. These are very complex hollow structures made from aluminum and silicon compounds. The hydrocarbons crack into smaller molecules as they are pumped through the zeolites.

BEST HYDROCARBONS FOR GASOLINE

The cracked hydrocarbons are then separated into fractions in the same way as before. Again only some of the products are useful as fuel. However, this time the unwanted molecules are too small and light to be used in gasoline.

The best hydrocarbons for gasoline are small, branched alkanes. They burn more slowly than unbranched compounds and keep an engine running smoothly.

Another process, called alkylation, converts the small, light alkanes and alkenes produced by cracking into larger branched alkanes. The catalysts used in alkylation are strong acids.

TAKING A CLOSER LOOK

ALMOST ORGANIC

Windshields of cars and other vehicles are coated with a thin layer of chain molecules of silicon atoms called silanes. The silanes stop rainwater from clinging to the glass so that the driver can see out better.

Carbon and hydrogen can form into any size of molecule. However, a couple of other elements—silicon and boron—can also form chains, too, only much shorter. Silicon and boron share certain properties with carbon. Silicon (Si) has four electrons in its outer shell, just like carbon. However, it has larger atoms, and they bond weakly with hydrogen. Nevertheless hydrogen and silicon atoms can form short chains that are similar to alkanes. These compounds are called silanes. Only monosilane (SiH_4) is very stable on its own. Boron (B) atoms are slightly smaller than carbon's. They have three outer electrons and can form only three bonds at once. Chains made from boron and hydrogen atoms are called boranes. The simplest is diborane (B_2H_6).

Oil wells are often in remote places. The crude oil is pumped through long pipes to refineries or to ports, where it is loaded onto a ship called a tanker, which then delivers the oil to a refinery.

Coal miners inspect some coal taken from a mine. Coal is mainly pure carbon, but that can be converted into useful hydrocarbons using catalysts such as cobalt and nickel.

ANOTHER SOURCE OF HYDROCARBONS

Crude oil is not the only source of hydrocarbons. Coal is a rock made from pure carbon mixed with hydrocarbons. Coal is used as a fuel and is reacted with ores to make pure metals. In the past coal was a source of a fuel called coal gas. This gas is poisonous and has been replaced with natural gas. However, coal may become a useful source of petrochemicals in the future.

TAKING A CLOSER LOOK

OIL SPILLS

Every day the world uses 85 million barrels of oil. One barrel of oil contains 42 gallons (159 liters). That adds up to 150 million gallons an hour! All this oil needs to reach refineries. Most of the oil is taken in giant ships called tankers.

The largest tankers carry 400,000 tons (360 metric tons) of oil. What happens when a tanker spills its oil? Crude oil floats on water, forming a "slick." Sometimes the gasoline and other fuels catch fire. If the slick can be kept together out at sea with long booms, it can be cleared up. Oil that washes onto the coast and into the mouths of rivers will kill fish, birds, and other wildlife and takes months to clear away.

Workers clean up an oil spill in Galicia after a tanker sank. The spill triggered extensive pollution along the coast.

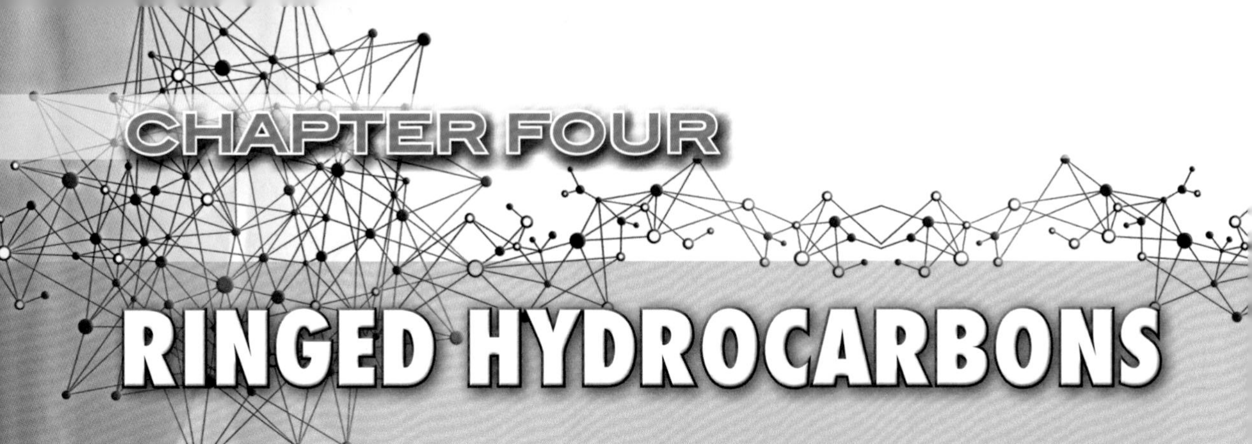

CHAPTER FOUR

RINGED HYDROCARBONS

As well as chains, hydrocarbon compounds can also form rings. Many of these ringed molecules have unusual properties.

Hydrocarbon molecules that form chains follow a structure that is similar to that of diamond. Like in a diamond crystal, the carbon and hydrogen atoms form into a series of pyramid structures. However, there are hydrocarbons with structures that are more similar to graphite.

KEY DEFINITIONS

- **Atom:** The smallest piece of an element that still retains the properties of that element.
- **Compound:** A substance formed when atoms of two or more different elements bond together.
- **Hydrocarbon:** A type of organic compound containing only carbon (C) and hydrogen (H) atoms.
- **Molecule:** Two or more atoms connected by chemical bonds.

Pieces of Styrofoam are made from a plastic called polystyrene that has been pumped full of air. Polystyrene is made up of many ringed molecules joined together.

Like diamond, graphite is a form of pure carbon. Instead of forming pyramids, the atoms in graphite form hexagons, or six-sided rings. Hydrocarbon molecules that contain similar hexagons are called arenes. Another name for them is aromatic compounds because many of them have a strong aroma (odor).

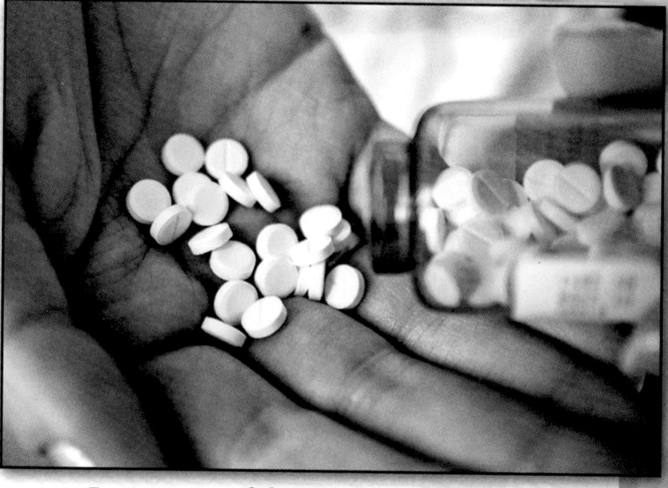

Aspirin, one of the most commonly used painkillers, is a compound called acetylsalicylic acid. This compound is an arene. It has molecules containing a ring of six carbon atoms.

THE SIMPLEST AROMATIC HYDROCABON

The simplest arene is a compound called benzene. A benzene molecule contains six carbon (C) atoms and six hydrogen (H) atoms. The compound's chemical formula is C_6H_6.

The six carbon atoms are connected into a hexagon. Each carbon atom is also bonded to a single hydrogen atom. Carbon atoms form a total of four bonds. In benzene, the carbon atoms are connected to just three other atoms. Thus, each carbon atom forms a double bond

TAKING A CLOSER LOOK

THE BENZENE RING

Benzene (C_6H_6) is the simplest arene. Its six carbon atoms form a ring. They are connected to each other by three single and three double bonds. The position of these bonds is not fixed. The double bonds and single bonds can swap places. As a result the molecule's three double bonds become shared between all six carbon atoms.

Benzene (C_6H_6)
double bond
carbon atom
hydrogen atom
single bond
=
benzene symbol
shared bond

Two ways of showing the structure of benzene.

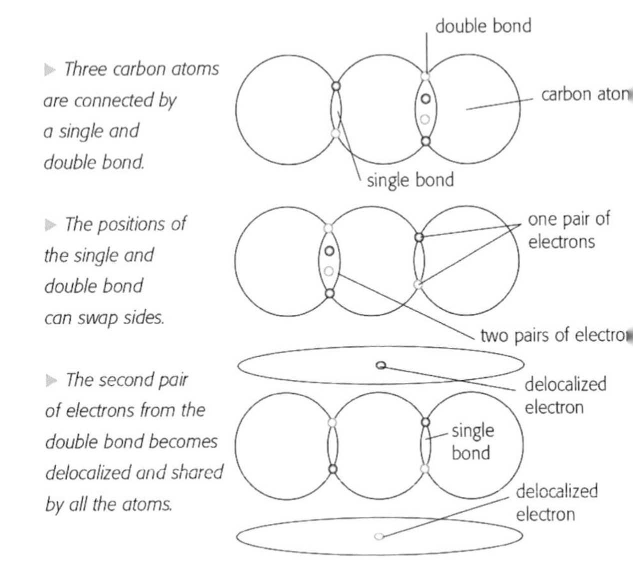

A diagram showing how the electrons in the double bonds in a benzene molecule become delocalized and shared between all the atoms.

with one of its neighboring carbon atoms. As a result, the hexagon of atoms is held together with a mixture of single and double bonds. The two types of bond are arranged alternately.

BENZENE RING

The bonds in a molecule of benzene are covalent. A covalent bond forms between atoms that are sharing electrons. The single bonds in the benzene molecule form when two carbon atoms share a pair of electrons. The double bonds form when two atoms share two pairs of electrons.

The two pairs of electrons in a double bond are not the same. The first pair forms in the same way as a pair in a single bond. The second pair of electrons are held together less strongly. They are more likely to break apart and form a bond with another atom.

In a benzene ring, each of the carbon atoms has formed a single and a double bond with the two carbon atoms on either side. Because the carbon atoms are connected in a ring, they all form a single bond on one side and a double bond on the other. For example, if one atom has a single bond on its left side and a double on the right, then all the other carbon atoms in the molecule will have those bonds in the same positions.

However, one pair of electrons can move from one side of the carbon atom to the other. As it does this, the double bond becomes a single bond, while the single bond becomes a double. As a result, that

TAKING A CLOSER LOOK

ARENE COMPOUNDS

An arene is any compound that contains one or more benzene-like rings in its molecule. Arene compounds may have a chain of carbon atoms branching out from the ring in place of a hydrogen atom. Other arene molecules contain two or more joined rings.

Some simple arenes.

Naphthalene $C_{10}H_8$

naphthalene symbol

$-CH_3$ (methyl)

Xylene $C_6H_4(CH_3)_2$

xylene symbol

Toluene $C_6H_4CH_3$

toluene symbol

Certain nylons, such as those in these ropes, are made from ringed hydrocarbons that are not arenes.

pair of electrons is effectively shared between the bonds on both sides of the atom. Because the carbon atoms form a ring, their double bonds fuse into a single shared bond.

The electrons inside this shared bond are described as delocalized. They are not linked to one bond but shared between the bonds, connecting several atoms. In benzene, the six delocalized electrons move around in doughnut-shaped spaces above and below the ring of carbon atoms.

Certain nylons, such as those in these ropes, are made from ringed hydrocarbons that are not arenes.

ROLE OF DELOCALIZED ELECTRONS

All arene compounds contain rings held together with delocalized electrons. Some arene compounds have a single ring with chained sections attached to them. Other arenes are made up of several rings joined together.

The delocalized electrons make the molecules of benzene and other arenes more stable than many other hydrocarbons. Chemists measure how strongly the atoms in a molecule are bonded to each other by measuring how much heat is released when the compound burns.

Burning is a reaction between a compound and oxygen (O_2). During the reaction, the compound breaks apart

TAKING A CLOSER LOOK
OTHER RINGS

Some ringed hydrocarbon molecules are not aromatic. The carbon atoms in these compounds may be connected by single bonds or have just one double bond. There are no delocalized electrons as in benzene molecules. Ringed compounds like this are called alicyclic compounds. Their chemical behavior is similar to that of alkanes or alkenes.

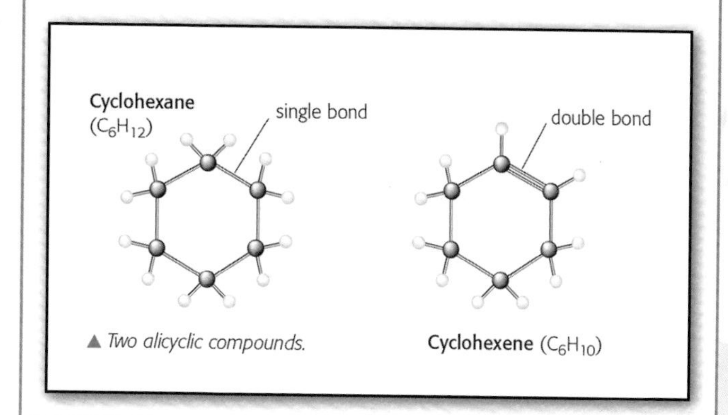

▲ *Two alicyclic compounds.*

KEY DEFINITIONS

- **Arene:** A type of hydrocarbon compound that has ringed molecules.
- **Aromatic:** Describes a compound that contains one or more benzene rings.
- **Benzene:** The simplest arene.
- **Covalent bond:** A bond in which two or more atoms share electrons.
- **Electron:** A tiny particle located inside an atom. Electrons are involved in forming bonds between atoms.

CHEMISTRY IN ACTION
MAKING A BANG

One of the most familiar and common explosives is an arene compound. Most people have heard of TNT. These letters stand for trinitrotoluene. TNT is used in bombs, to demolish old buildings, and to blast away rocks in mines.

Toluene is an arene compound. It is a benzene ring with a methyl group ($-CH_3$) attached. A molecule of TNT is a toluene molecule that also has three nitro groups ($-NO_2$) attached.

When it explodes, TNT is very powerful. However, it is also relatively stable. It does not react easily and stays safe when it gets hot or wet. TNT will explode if a detonator is used to make it hot enough to react; that is 563 degrees Fahrenheit (563°F; 295°C). At this temperature the molecules of TNT break apart, and the nitro groups react with each other to make gases. These gases expand very rapidly and cause a shockwave in the air. It is this shockwave that causes damage to solid objects in its path.

The power of large explosions is measured in kilotons. A 1-kiloton explosion releases the same amount of energy as an explosion of 1,000 tons of TNT.

A U.S. Navy student detonates a standard TNT charge during SEAL training in basic underwater demolition tactics.

and its atoms bond with oxygen atoms. The heat released during burning is the energy left over after the old molecules have broken apart and formed into new ones. More stable hydrocarbons release less heat when they burn. That is because more of the energy has been used up breaking the strong bonds in their molecules.

Chemists can measure how strong a single bond between two carbon atoms is by burning simple alkanes. They can also test the strength of a double bond by burning an alkene. Chemists could then add these values together to figure out how strong the three single and three double bonds are in a ring of benzene.

However, when this result is tested by burning benzene in a laboratory, chemists find that the molecule's bonds are stronger than they thought. The delocalized electrons are shared equally among all the bonds, making them all stronger.

REACTIONS OF ARENES

The stability of the delocalized electrons has an effect on the way benzene and other arenes react. Hydrocarbons with double bonds in their molecules tend to be reactive. That is because the double bond easily breaks open and forms two single bonds. For example,

PROFILE
DISCOVERING BENZENE

Michael Faraday is one of the most important scientists in history. He discovered many things about electricity and magnetism. With that understanding, he invented the first electricity generator and the first electric motor. However, one of Faraday's first discoveries is less well known. For it was Faraday who discovered benzene.

Michael Faraday poses with some of his laboratory equipment in the early 1860s.

Michael Faraday was born close to London, England, in 1791. The young Michael often went hungry and was not able to get a good education. Nevertheless, at the age of 14 Faraday was already experimenting, using home-made batteries to investigate electricity.

At the age of 20, Faraday became the apprentice (assistant) of Humphry Davy (1778–1829). At that time, Davy was making many discoveries. He had used electricity to purify several then-unknown elements, including the metals sodium and potassium.

Faraday completed his service with Davy in 1820. At first he studied chemistry, only turning to physics ten years later. In 1820, Faraday produced the first chlorocarbons (compounds of chlorine and carbon). In 1825, he described a liquid compound, which he named bicarbuet of hydrogen. A few years later, the German chemist Eilhard Mitscherlich (1794–1863) renamed this compound benzene after making it from benzoin resin.

an alkene will react with hydrogen (H_2) to become an alkane. In this reaction, the alkene's double bond breaks and forms single bonds with hydrogen atoms.

That sort of reaction is called an addition because the hydrogen atoms are added to the molecule. With three double bonds in its molecule, you might expect benzene also to be reactive in that way. However, benzene and other arenes do not take part in addition reactions with hydrogen. The delocalized electrons stop the molecule's double bonds from breaking and forming into single bonds.

A liquid pesticide is sprayed onto a field to kill insects that attack crop plants. Many pesticides contain arene compounds.

CHEMISTRY IN ACTION
AZO DYES

Many of the dyes used to color our clothes are aromatic compounds. These compounds are called azo dyes because they contain a section in the molecule called an azo group. An azo group is formed when one part of a hydrocarbon is connected to another section by two nitrogen (N) atoms. The nitrogen atoms sit between the two parts of the molecule and are connected to each other by a double bond.

Azo groups can form in chained hydrocarbons, but these compounds are very unstable. However, when an azo group attaches to a benzene ring, it forms a stable molecule. The double bond between the nitrogen atoms becomes part of the benzene's system of delocalized electrons. That keeps the molecule stable.

Many azo compounds are brightly colored, mostly red, orange, or yellow. Azo dyes were first used in the 1880s. The first was called Congo red. However, this and other early dyes have been replaced with other azo dyes, which last longer. Most azo dyes are poisonous, but some, such as tartrazine yellow, are used to color food.

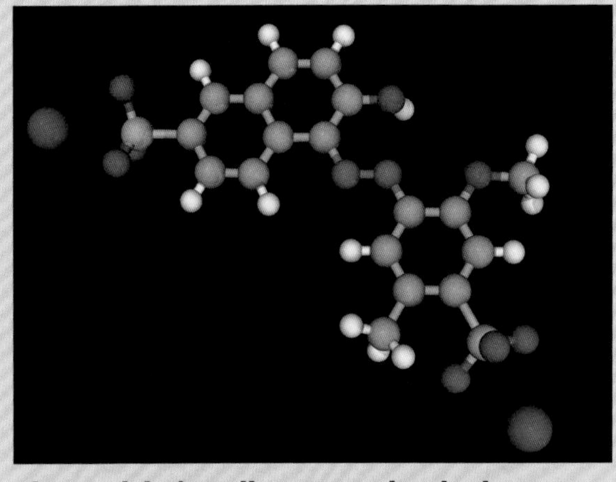

This model of an allura AC molecule shows red azo dye used as food dye. Atoms are represented as spheres and color coded: carbon as gray, hydrogen as white, nitrogen as blue, oxygen as red, and sulfur as yellow. The sodium ions are violet.

CHEMISTRY IN ACTION
MOTHBALLS

Have you heard the expression "put into mothballs?" People say it to mean something is being stored for a long time. The expression comes from the way people sometimes add mothballs to clothes before packing them away for a long time. The small balls are made from naphthalene, an arene compound. The smell of the balls keeps moths away. Moth caterpillars would otherwise eat the wool and cotton in the clothes and make holes in them.

Moths are eating this wool sweater. The caterpillars of clothes moths love to eat natural fibers such as wool.

Instead, arenes undergo displacement reactions, where an atom or molecule takes the place of one of the hydrogen atoms. Only very reactive elements, such as the halogens, will react with arenes in this way. For example, chlorine (Cl_2) reacts with benzene (C_6H_6) to make chlorobenzene (C_6H_5Cl) and hydrochloric acid (HCl). The chemical equation for this reaction is:

$$C_6H_6 + Cl_2 \rightarrow C_6H_5Cl + HCl$$

A chemist holds a solution of fullerene dissolved in benzene. Fullerenes are a type of pure carbon found in soot. Like benzene and other arenes, fullerenes have delocalized electrons.

BENZENE POISONING AND ARENE MEDICINES

Benzene and several other arene compounds are very poisonous. Even a tiny amount of benzene in food or water is enough to make a person ill. Benzene damages the body's immune system and nerves and causes cancer. Many pesticides (compounds used to kill pests) are arenes. However, many life-saving medicines and painkillers are also arene compounds.

CHAPTER FIVE

FUNCTIONAL GROUPS OF ALCOHOLS AND ORGANIC ACIDS

Glasses of red and white white. Wine is an alcoholic beverage made from grape juice. The sugar in the juice reacts with oxygen to make alcohol.

Not all organic compounds are hydrocarbons. Many of them contain atoms of other elements. Oxygen is a common ingredient in organic compounds. These compounds include alcohols and organic acids, such as vinegar.

Hydrocarbons are compounds that contain just carbon (C) and hydrogen (H) atoms. As we have seen, the bonds between atoms of these elements are very strong and unlikely to break. As a result, hydrocarbons are not very reactive. However, when atoms of other elements are added to the compounds, they become more reactive. That is because the bonds between the carbon atoms and these other atoms are much weaker and more likely to break open and take part in a reaction.

TAKING A CLOSER LOOK
ALCOHOLS

When a hydroxyl group (-OH) containing an oxygen and hydrogen atom joins to a hydrocarbon chain, it forms a compound called an alcohol. Alcohols are named according to the number of carbons in their molecules. Their names generally end in -ol.

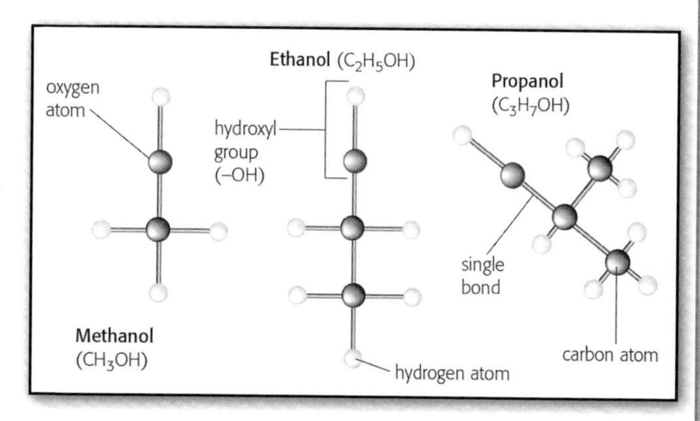

Three simple alcohol compounds.

The section of an organic molecule that contains an atom that is not carbon or hydrogen is called a functional group. The structure of a functional group determines how that compound will react with other chemicals.

BONDING WITH OXYGEN

Oxygen (O) forms a number of functional groups in organic compounds. An oxygen atom can form a total of two bonds with other atoms. For example, water is a compound of oxygen and hydrogen. One oxygen atom is bonded to two hydrogen atoms to make the molecule H_2O. Imagine if one of the hydrogen atoms was replaced with a carbon atom in a hydrocarbon molecule. That hydrocarbon would then have an oxygen and hydrogen (-OH) attached to it. This -OH structure is a functional group called a hydroxyl. Organic chain compounds with a hydroxyl group belong to the group of compounds called the alcohols. Hydroxyl groups on an arene compound form compounds called phenols.

KEY DEFINITIONS

- **Compound:** A substance formed when atoms of two or more different elements bond.
- **Hydrocarbon:** An organic compound containing only carbon and hydrogen atoms.
- **Functional group:** A part of an organic molecule that gives it certain chemical properties.
- **Molecule:** Two or more atoms connected together.
- **Organic:** Describes a compound that contains carbon and generally hydrogen but also atoms of other elements.

MANUFACTURING ALCOHOLS

Alcohols are some of the most familiar organic compounds. They occur naturally, and people have been making them for thousands of years. The most common alcohol compound is ethanol (C_2H_5OH). Ethanol is often called grain alcohol because it can be made from sugars in grain and fruits. That occurs through a series of reactions called fermentation.

Fermentation reactions involve sugar reacting with oxygen. They occur in nature, and they are used to make alcoholic beverages.

The other common type of alcohol is methanol (CH_3OH). This compound is sometimes called wood alcohol because it can be made by heating wood. If the wood is kept out of the air so it does not burn, it will produce methanol vapors.

Methanol is poisonous, just like all alcohols. Only ethanol can be consumed in small quantities. But in large amounts even ethanol can kill. Most of the simple alcohols are used as solvents.

Alcohol molecules that have two hydroxyl group in their molecules are

KEY DEFINITIONS

- **Electron:** A tiny particle that makes up part of an atom. Electrons have a negative charge.
- **Fermentation:** A reaction in which sugar is turned into ethanol.
- **Hydroxyl:** A functional group made up of an oxygen and a hydrogen atom.

Beer is made by fermenting grains and hops.

CHEMISTRY IN ACTION
FERMENTATION

Alcoholic drinks, such as wine and beer, are made using a natural process called fermentation. Ethanol (C_2H_5OH) is produced by fermenting glucose, a sugar ($C_6H_{12}O_6$). The reaction also produces carbon dioxide (CO_2). Fermentation occurs inside living cells. The reaction releases energy. The equation for the reaction is:

$$C_6H_{12}O_6 \rightarrow 2C_2H_5OH + 2CO_2$$

The alcohol in wine and beer is fermented by yeast cells. For wine, the sugar comes from grapes. For beer, it comes from grains. The yeast turns the sugars into ethanol until the amount of alcohol gets so high, it kills the yeast. As a result, drinks made like this are not more than about 12 percent ethanol. Strongly alcoholic drinks, such as whiskey, are made by purifying the ethanol.

TAKING A CLOSER LOOK

ISOMERS

Alcohol compounds with three or more carbon atoms form isomers. Isomers are molecules that contain the same atoms, but those atoms are arranged in different ways. Isomers have the same formula, but their molecules have a different shape. Those shapes may create different functional groups and so effect the molecule's behavior.

For example, there are three isomers with the formula C_3H_8O. Two of them are types of propanol (C_3H_7OH) because they each have a hydroxl group (-OH). The two molecules are called propan-1-ol and propan-2-ol (rubbing alcohol). The number tells us which carbon atom has the hydroxyl group attached.

The third isomer is not an alcohol. Instead of forming a hydroxyl group, the oxygen atom joins two carbon atoms together. A molecule with this functional group is called an ether. This ether is called methoxyethane. Ethers react in different ways to alcohols.

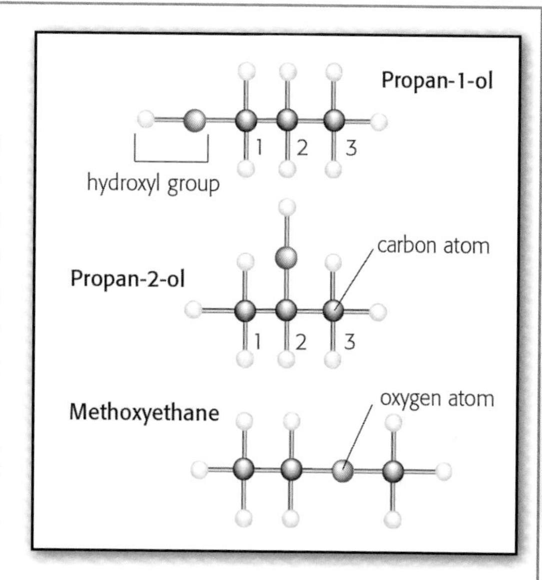

The three isomers of C_3H_8O. Numbers 1–3 refer to the position of the carbon atoms.

called glycols. A few compounds have even more hydroxyl groups. For example, glycerol—$C_3H_5(OH)_3$—has three hydroxyl groups.

FORMING HYDROGEN BONDS

Oxygen atoms are very reactive. They pull on electrons more strongly than do the atoms of most other elements. The oxygen atom in an alcohol molecule pulls electrons away from the carbon and hydrogen atoms nearby. As a result, the oxygen atom becomes slightly negatively charged. The hydrogen bonded to it has a slight positive charge. Opposite

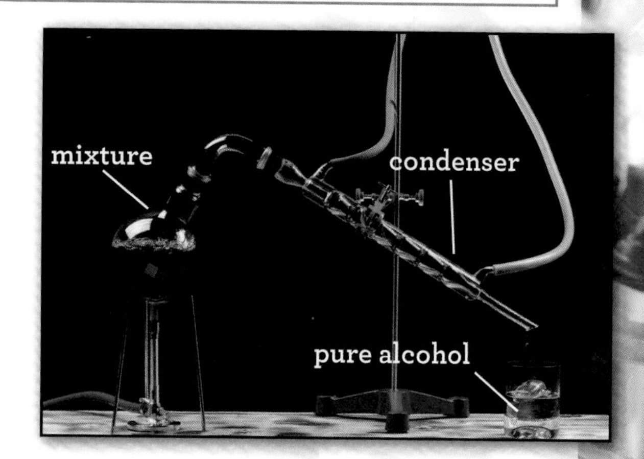

Pure alcohol is made by distillation. In this process, a mixture containing alcohol is heated so that the alcohol boils. The vapor is then turned back into a liquid by a water-cooled condenser.

A CLOSER LOOK
PHENOL

When a hydroxyl group (-OH) bonds to a benzene ring, it forms a special alcohol. The compound is called a phenol. The simplest phenol is C_6H_5OH.

In a benzene ring, some electrons are spread evenly around the molecule. That makes a hydroxyl group behave differently in a phenol from how it does in an alcohol. The ring of shared electrons expands to include the oxygen atom (O) and holds it very tightly to the molecule. The bond between the oxygen atom and its hydrogen atom (H) becomes weak. The bond breaks easily. That produces two ions: H+ and C_6H_5O-. Compounds that split up in this way are called acids. Phenols are weak acids.

Phenol crystals have a pink color. They dissolve in water to make an acid called carbolic acid.

charges attract each other, so the hydrogen atom on one alcohol molecule is drawn toward the oxygen atom on the other. That creates a weak bond between the two molecules. This attraction is called a hydrogen bond. Many oxygen compounds form hydrogen bonds, including water.

Hydrogen bonds hold alcohol molecules together more strongly. As a result, the boiling points of alcohols are higher than those of hydrocarbons without oxygen. The boiling point is the temperature at which the molecules in a liquid break free of each other and turn into a gas. Both methanol and ethanol are liquids in normal conditions. Without hydrogen bonds between their molecules, these alcohols would be gases.

The negative charge of the oxygen atom also affects the way alcohols react. For example, they react with oxygen to make compounds called aldehydes and ketones. Alcohols will also react with

When it is mixed with water, phenol becomes carbolic acid. In the past, that was used in strong soaps.

CHEMISTRY IN ACTION
ANTISEPTICS

Today, a surgeon's operating room is very clean. No one is allowed in unless they wash themselves and put on protective clothing. If any dirt got into a patient's body during an operation, he or she might become very ill and die. The dirt contains tiny life-forms called bacteria, which infect the body and cause illness.

About 150 years ago, people did not understand these risks. People often died after operations, not because of the surgery itself but because of infections. In 1865, a British surgeon named Joseph Lister (1827–1912) began using carbolic acid (phenol in water) to make operating rooms sterile (free of bacteria). The phenol was acidic enough to kill bacteria, but it was also mild so that it did not hurt the patient. Lister's idea led to the way surgery is performed today.

Joseph Lister used this carbolic steam spray around 1867. The carbolic acid (phenol and water) spray antiseptic system allowed him to carry out surgery without patients developing deadly infections. He soaked his surgical gauzes and instruments in carbolic acid.

oxygen to form acidic compounds called carboxylic acids.

CARBOXYL FUNCTIONAL GROUP

Carboxylic acids have two functional groups. One of them is a hydroxyl (-OH), the group also found in alcohols and phenols. The other functional group is called a carbonyl (-CO). In this group, an oxygen atom is bonded to a carbon atom. The two atoms are connected by a double bond.

In a carboxylic acid molecule, both of these functional groups are attached to the same carbon atom. Together they form a carboxyl group (-COOH).

NATURAL CARBOXYLIC ACIDS

Just like the alcohols and other organic compounds, carboxylic acids are named

Lime juice contains citric acid, a type of carboxylic acid. The citric acid is what gives limes their sharp taste.

TAKING A CLOSER LOOK
CARBOXYLIC ACIDS

The main group of organic acids is called the carboxylic acids. They have a hydroxyl (-OH) and carbonyl (-CO) group bonded together. Some carboxylic acids have several of these groups.

Carboxylic acids split into ions. A hydrogen ion (H+) breaks away from the rest of the molecule, which becomes a negatively charged carboxylate ion. For example, methanoic acid forms methanoate ions (HCOO-).

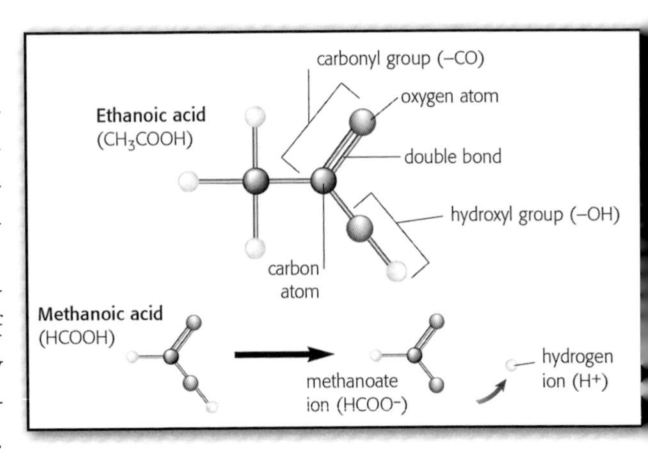

according to how many carbon atoms there are in their molecules. All their names end in -oic.

However, many carboxylic acids are found in foods or occur elsewhere in nature, and over the years they have been given other names. For example, vinegar contains ethanoic acid (CH_3COOH). This compound is also known as acetic acid. Methanoic acid (HCOOH) is also known as formic acid.

There are many other natural carboxylic acids. These include citric acid from lemons, oranges, and other citrus fruits. Lactic acid is made by muscles when they work hard. It is this acid reacting with other compounds in the muscles that makes them ache and feel tired.

Longer carboxylic acid molecules are called fatty acids. They are found in milk, oils, and fats. For example, lauric acid occurs in coconut milk.

REACTIVE ACIDS

Carboxylic acids are made when alcohols react with oxygen (O_2). Ethanoic acid is made in nature as part of the same fermentation process that turns sugars into ethanol (C_2H_5OH). As we have seen, yeast makes the alcohol, but the changes do not end there. If the ethanol is exposed to the air, bacteria mixed into it will turn

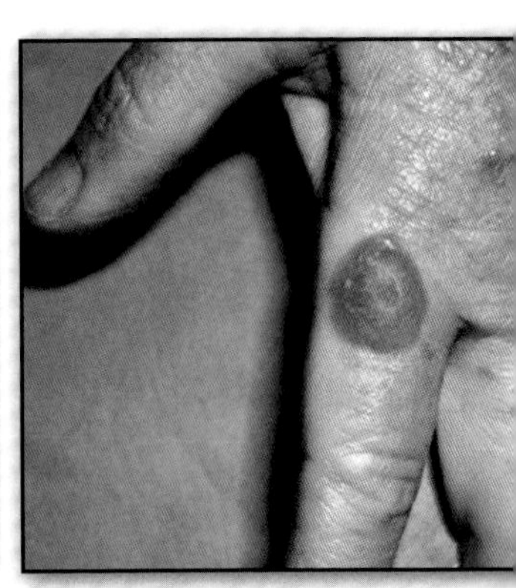

The stings of fire ants contain formic acid (HCOOH). The acid causes painful blisters on the skin.

CHEMISTRY IN ACTION

PICKLING

Food can be stored for a long time by pickling. Food is pickled in either vinegar (ethanoic acid) or strong alcohol (ethanol). The acid in vinegar stops bacteria on the food from growing. Food that has bacteria growing on it will go bad. The vinegar also soaks into the food, giving it a strong flavor. Pickling in alcohol also gives the food a certain flavor. However, this time the food is preserved thanks to bacteria, which slowly turn the ethanol into ethanoic acid. This acid then stops the food from going bad.

These pickles are small cucumbers that have been preserved in salty vinegar.

it into ethanoic acid. This reaction has the following equation:

$$C_2H_5OH + O_2 \rightarrow CH_3COOH + H_2O$$

That is why wine and other alcoholic beverages will begin to taste sour after being opened for a long time. They are slowly turning into vinegar!

Positively charged hydrogen ions (H+) break off carboxylic acid molecules. That is why they are classed as acids. Acids are reactive compounds because they produce hydrogen ions. Acids react with other compounds to produce substances called salts.

When it loses its hydrogen ion, the rest of the carboxylic acid becomes a negatively charged ion. During a reaction, this ion forms a salt. The salts of carboxylic acids have names ending in -oate. When ethanoic acid (CH_3COOH) reacts with calcium hydroxide $Ca(OH)_2$, it produces calcium ethanoate, $Ca(CH_3COO)_2$, and water (H_2O). The equation for this reaction looks like this:

$$2CH_3COOH + Ca(OH)_2 \rightarrow Ca(CH_3COO)_2 + H_2O$$

KEY DEFINITIONS

- **Acid:** A compound that splits easily into a positively charged hydrogen ion and another negatively charged ion.
- **Benzene:** A ring of carbon atoms in which some electrons are shared by all atoms in the molecule.
- **Carbonyl:** A functional group in which an oxygen atom is connected to a carbon atom by a double bond.
- **Ion:** An atom or molecule that has lost or gained one or more electrons and has become electrically charged.

CHAPTER SIX

DIFFERENT ORGANIC FUNCTIONAL GROUPS

There are many types of functional groups, and each one gives organic compounds certain properties. As well as containing oxygen atoms, there are also functional groups that contain atoms of other elements.

There are many classes of organic compounds. A compound is formed when atoms of two or more elements bond together. Organic compounds are made up of mostly atoms of carbon (C) and hydrogen (H). Compounds containing only these elements are known as hydrocarbons. However, many organic molecules also contain atoms of other elements. Where these atoms bond to the hydrocarbon, they form a functional group. The functional group has an effect on the way that compound behaves. Organic compounds are classed according to

KEY DEFINITIONS

- **Compound:** Atoms of different elements bonded together.
- **Functional group:** A section of an organic molecule that gives it certain chemical properties.
- **Molecule:** Two or more atoms connected together.
- **Organic:** Describes a compound that is made of carbon and that generally also contains hydrogen.

Many of nature's smells and tastes, such as the fragrance of flowers, are produced by organic compounds. Pleasant smells are from compounds containing oxygen, while foul-smelling compounds contain nitrogen or sulfur.

TAKING A CLOSER LOOK
ESTERS

When alcohols react with carboxylic acids, they form compounds called esters. An ester molecule has two halves, one side coming from the alcohol and the other from the acid. These two sections are connected by an oxygen atom. The side of the molecule that was originally from the acid contains a carbonyl group (-CO).

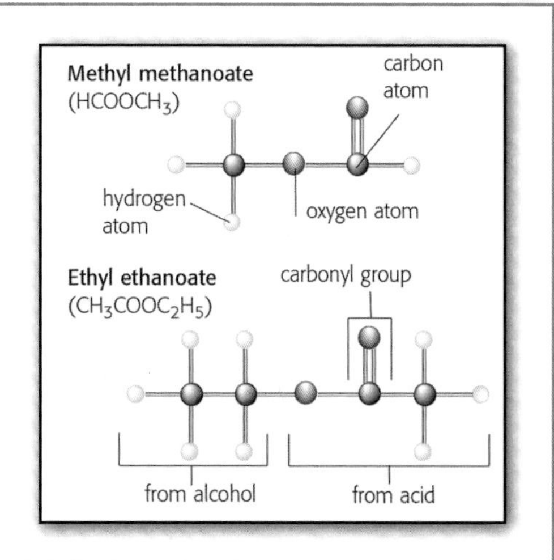

Two simple ester molecules.

their functional groups. There are many functional groups. This chapter looks at some of the groups containing oxygen atoms (O), nitrogen (N), sulfur (S), and chlorine (Cl).

ALCOHOL AND CARBOXYLIC ACID

As we have seen, alcohol and carboxylic acids are types of organic compounds that are common in the natural world. They both have functional groups containing oxygen.

When an alcohol and carboxylic acid react, they form a compound called an ester. An ester forms when an alcohol's functional group reacts with the functional group of the acid. Alcohols have a hydroxyl functional group made from an oxygen and hydrogen atom. Carboxylic acids have a hydroxyl group, too, but they also have a carbonyl group. That is a carbon atom connected to an oxygen atom by a double bond.

To make an ester, the alcohol loses a hydrogen atom from its hydroxyl group. The oxygen atom left behind attaches to the carbon atom in the acid's carbonyl group. The acid also loses its hydroxyl to form an ester. The hydroxyl group and hydrogen atom also bond together to make water (H_2O).

Like all organic compounds, esters are named according to how many carbon atoms they have. The simplest ester is methyl methanoate ($HCOOCH_3$). This

The smell of pineapples is produced by an ester called ethyl butanoate.

CHEMISTRY IN ACTION
FROM FATS TO SOAPS

Vegetable oils and animal fats are complex ester compounds called triglycerides. They are formed when three large carboxylic acids bond to an alcohol called glycerol. The carboxylic acids in fats and oils are long-chained molecules. They are called fatty acids.

Soap is made from triglycerides. An alkali, such as sodium hydroxide (NaOH), is added to the esters. That makes each ester split into a glycerol and three fatty acids. The sodium forms compounds with the acids. These compounds are waxy solids and are filtered out and dried. Perfume and coloring is then added before the soap is pressed in bars.

Soap is made from the ester compounds in fats and oils.

compound is produced when methanol (CH_3OH) reacts with methanoic acid (HCOOH; also called formic acid). The equation for the reaction is:

$$HCOOH + CH_3OH \rightarrow HCOOCH_3 + H_2O$$

The ester's name has two parts because its molecule is in two parts. The ester is named methyl methanoate. One side comes from the methanol. That has lost a hydrogen atom and becomes the methyl part of the ester. The section from the methanoic acid has lost its hydroxyl group. The section produced by that is called methanoate.

SMELLS AND FATTY ACIDS

Most small ester compounds are liquids. These liquids evaporate easily (turn into a gas), and many of them have a distinctive smell. For example, the smell of a banana is an ester called isopentyl ethanoate. Many of the artificial flavors used in candy are esters. They are also added to perfumes.

A molecule of fat. The molecule has three fatty acids attached to a central glycerol molecule. Together they form a complex ester. The red spheres are the oxygen atoms holding the molecule together.

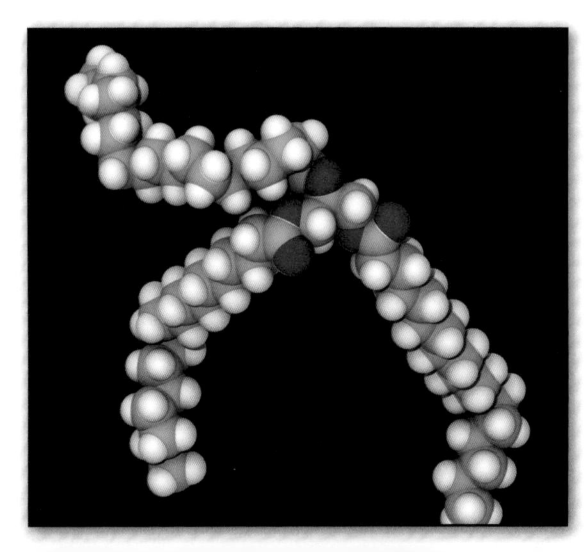

TAKING A CLOSER LOOK

ALDEHYDES

There are two types of compounds that have just a carbonyl as their functional group. Aldehydes have this carbonyl at the end of each molecule. Ketones have carbonyls in the middle of each molecule. The carbonyl group is made up of an oxygen connected to a carbon atom by a double bond. It is a very reactive functional group.

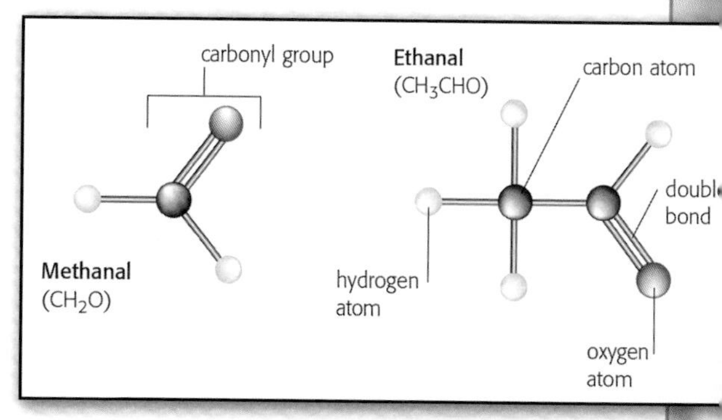

The two simplest aldehyde compounds.

A dead bird is kept preserved in a jar of water and methanol, a type of aldehyde. Methanol, also known as formaldehyde, prevents the bird's body from decaying.

Large ester compounds do not evaporate easily. Instead they are oily liquids and waxy solids. Some animal fats and vegetable oils are complex ester compounds. Such fats and oils are made up of three carboxylic acid molecules connected to an alcohol called glycerol. Glycerol molecules have three hydroxyl groups (-OH), and each one forms an ester with a large carboxylic acid, known as a fatty acid. Fatty acids may be saturated or unsaturated. Saturated molecules contain only single bonds, while unsaturated molecules have one or more double bonds. Most animal fats are saturated compounds; many vegetable oils are unsaturated fatty acids.

KEY DEFINITIONS

- **Aldehyde:** A compound with a carbonyl group attached to the end of its molecule.
- **Carbonyl:** A functional group made by a carbon atom connected to an oxygen atom by a double bond.
- **Ester:** A compound formed when an alcohol reacts with a carboxylic acid.
- **Evaporate:** To turn from liquid to gas.

CARBON ATOMS, ALDEHYDES, AND KETONES

There are two classes of organic compounds that have a single carbonyl

(-CO) as their functional group. They are the aldehydes and ketones. Both classes of compounds are very similar. The only difference in their structures is that an aldehyde has a carbonyl group attached to the end of its molecule. A ketone has the same group attached in the middle of the molecule. These compounds are classed as two groups because the difference in their structure has an effect on their reactivity.

The names of aldehydes and ketones are based on how many carbon atoms they contain. Aldehydes have names that end in -al, while ketones have names ending in -one.

The simplest aldehyde is methanal (H_2CO). This compound is perhaps more familiar by its old name, formaldehyde. The simplest ketone is propanone (CH_3COCH_3). Again, this compound is also commonly referred to by its older name, acetone.

ALDEHYDES AND KETONES

Methanal is a volatile liquid. A volatile substance is one that changes into a gas easily. Methanal has a boiling point of just 69 degrees Fahrenheit (69°F; 21°C). That is about normal room temperature. On a hot day, the compound boils away! Methanal has a very sharp and unpleasant odor.

Propanone is also a liquid. It boils at 133°F (56°C). Like other ketones, propanone has a sweet smell.

TAKING A CLOSER LOOK

KETONES

The simplest ketone has three carbon atoms. That is because a ketone has to have its carbonyl group located in the middle of the molecule. Apart from propanone, other ketone molecules have numbers in their names. These numbers show which carbon atom the carbonyl group is attached to.

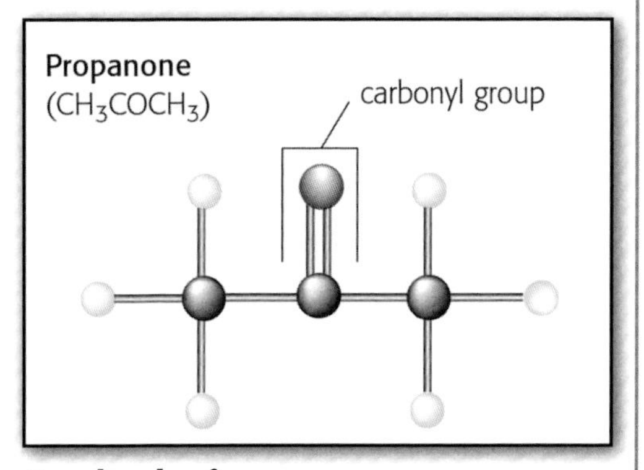

A molecule of propanone.

KEY DEFINITIONS

- **Boiling point:** The temperature at which a liquid turns into a gas.
- **Ether:** Compound in which two hydrocarbon groups are joined by an oxygen atom.
- **Ketone:** A compound with a carbonyl group attached in the middle of its molecule.
- **Volatile:** Describes a liquid that evaporates (turns into a gas) easily.

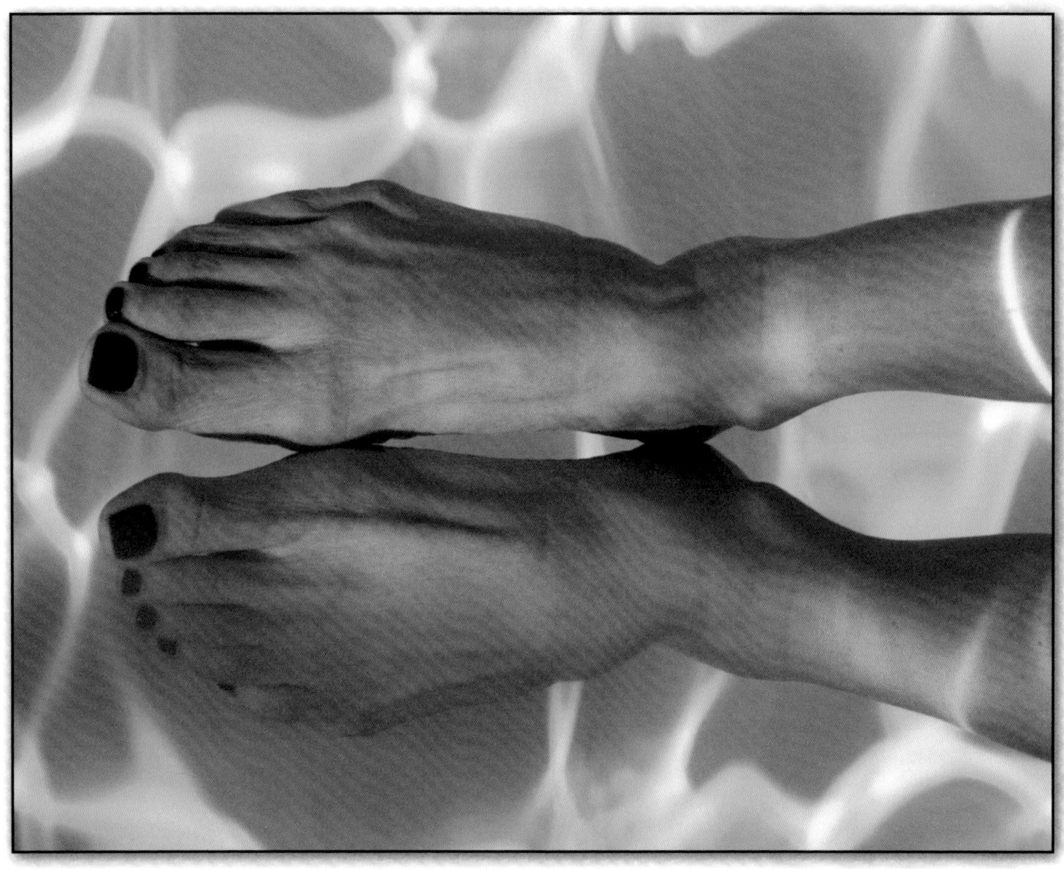

Liquid ketones such as propanone are used as solvents—liquids that dissolve other compounds. For example, nail polish is waterproof and does not wash off in water. Nail polish remover contains ketones, which can dissolve the nail polish and wash it away.

REACTIVITY OF CARBONYLS

Aldehydes and ketones are more reactive than most organic compounds because of the carbonyl group. The double bond holding the oxygen to the compound is very likely to break and form two single bonds. The oxygen atom pulls the electrons away from the carbon atom and gains a slight negative charge. Other molecules are attracted to this charge, and that is why the oxygen atom is so likely to be involved in a reaction.

An aldehyde is more reactive than a ketone because its carbonyl group is exposed at the end of the molecule. There it is bonded to one carbon atom and one hydrogen atom. As a result, the electrons are pulled very close to the oxygen atom. In a ketone, the carbonyl group is bonded between two carbon atoms. These larger atoms stop the oxygen atom from pulling the electrons quite so close. As a result, the aldehyde's oxygen atom has a stronger charge than a ketone's, and the aldehyde is more likely to react.

Aldehydes and ketones are halfway between alcohols and carboxylic acids. An alcohol can become an aldehyde or ketone by losing two hydrogen atoms. For example, methanol (CH_3OH) becomes methanal (H_2CO) through the reaction:

$$CH_3OH \rightarrow H_2CO + H_2$$

Methanal can be turned into methanoic acid (HCOOH) by adding oxygen (O_2). That reaction is as follows:

$$2H_2CO + O_2 \rightarrow 2HCOOH$$

ORGANIC COMPOUNDS CALLED ETHERS

An organic compound that has an oxygen atom bonded between two carbon atoms is called an ether. Ethers are made when two alcohols join together. During this reaction, two hydrogen atoms (H) and an oxygen atom (O) are taken from the alcohol molecules and become a molecule of water (H_2O). The reaction is described as a dehydration because water is being

CHEMISTRY IN ACTION
MAKING PERFUMES

Several familiar odors and flavors are produced by ketone compounds. For example, the taste of cheese comes from complex ketones. Musk is another fragrant ketone. It is used to make expensive perfumes. Its odor comes from the compound muscone. Musk is used in perfumes because other fragrances, such as esters, can be mixed into it easily to make pleasant odors.

Natural musk is produced by a small type of deer. For many years, musk deer were hunted for their musk glands. Today, the deer are very rare, and muscone is made in laboratories.

Musk comes from a gland on the belly of a male musk deer. Males use it to mark their territory. Deer are sometimes killed for their musk.

Many perfumes contain musk. Other fragrant chemicals are mixed into it.

Methoxymethane is used to make the spray in aerosol cans.

removed from the alcohols as they are joined together.

The simplest ether compound is called methoxymethane (CH_3OCH_3). The molecule is named for the two sections on either side of the oxygen atom. Methoxymethane is made by dehydrating two molecules of methanol (CH_3OH). The reaction is as follows:

$$CH_3OH + CH_3OH \rightarrow CH_3OCH_3 + H_2O$$

Because their oxygen atoms are bonded strongly to two carbons atoms,

ETHERS

Ether molecules have two halves connected to each other by an oxygen atom. The molecules are named for these two sections. The last part of the name relates to the size and structure of the largest section. That section takes the name of its equivalent hydrocarbon. For example, if the section has two carbon atoms, it is called ethane. The smaller section's name is also based on its number of carbon atoms. However, an –oxy is added to show that it is connected to the larger section by an oxygen atom.

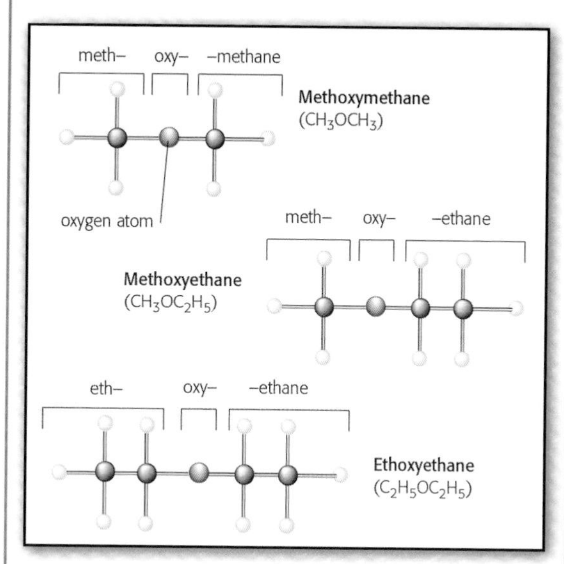

Three simple ether compounds.

The flavor of aniseed is produced by a complex ether called anethole. The ether also flavors fennel and licorice.

CHEMISTRY IN ACTION
SMELLY SULFUR

Sulfur is in the same group as oxygen in the periodic table. Therefore, the two elements share similar properties. Like oxygen atoms (O), sulfur atoms (S) can form a total of two bonds. Sulfur forms organic compounds that are similar in structure to those formed by oxygen.

For example, thiols are sulfur's equivalent to alcohols. Instead of having an –OH group, thiols have an –SH group. Thiols have a rotten-egg odor. For example, the smell produced by a skunk to repel attackers is butane-1-thiol (C_4H_9SH). Tiny amounts of that thiol are also added to natural gas to give it a distinctive smell.

ethers are not very reactive. However, ethoxyethane ($C_2H_5OC_2H_5$) was used as the first anesthetic.

AMINES

An organic compound that contains a single nitrogen atom (N) is called an amine. Nitrogen atoms can form three bonds. In an amine, all of them are single bonds.

The nitrogen atom sits at the center of the molecule, which is in three sections. In a simple amine, two of those sections might simply be hydrogen atoms. At least one of the sections is an alkyl group. An alkyl group is a

The distinctive smell of fish is from an amine compound called trimethylamine.

TAKING A CLOSER LOOK

AMINES

An amine is an organic compound containing a nitrogen atom. Amines are similar to ammonia. An amine compound has hydrocarbons in place of at least one of the hydrogen atoms. Amines are reactive compounds. They are used to make certain dyes.

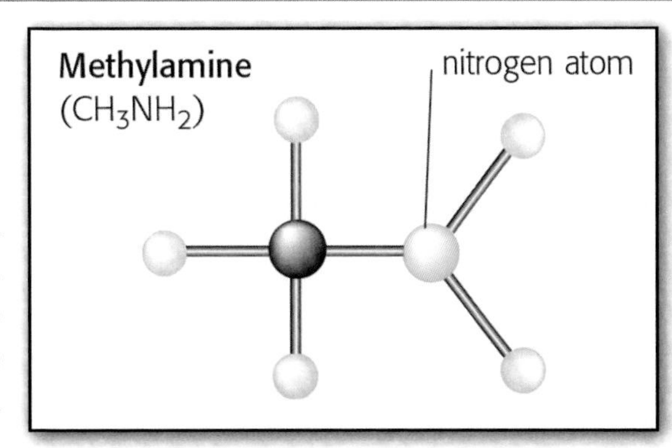

A molecule of methylamine, the simplest amine compound.

section of hydrocarbon that branches off another molecule or is attached to a functional group. For example, the simplest alkyl group is a methyl ($-CH_3$). So the simplest amine is called methylamine (CH_3NH_2).

The nitrogen atom may be bonded to two alkyl groups, for example, a methyl and an ethyl ($-C_2H_5$). In this case the alkyl groups are listed alphabetically, so the molecule is named ethylmethylamine. When there are two methyl groups, it is dimethylamine. When there are three, then the name is trimethylamine.

KEY DEFINITIONS

- **Alkyl group:** A section of hydrocarbon chain attached to a functional group.
- **Amine:** An organic compound containing a single nitrogen atom.
- **Halide:** A compound containing a halogen atom, such as chlorine or iodine.

HALOGEN ATOMS

The halogens are a very reactive group of elements. They include fluorine (F), chlorine (Cl), and bromine (Br). These elements form compounds called organic halides.

In an organic halide, one or more halogen atoms is bonded to a carbon atom. The halogens take the place of the hydrogen atom, and in some organic halides there are no hydrogen atoms at all. The halides are named for the halogens in them. Bromocarbons contain bromine, while chlorofluorocarbons contain both chlorine and fluorine atoms.

In many ways organic halides are very similar to hydrocarbons. They have slightly higher melting and boiling points than corresponding hydrocarbons. The halogen atoms are much heavier than hydrogen atoms, so the molecules need to be hotter before they break apart.

TAKING A CLOSER LOOK
CHLOROMETHANES

Chlorine forms four chloromethanes. These compounds have between one and four chlorine atoms (Cl) bonded to a single carbon atom. Chloromethane (CH_3Cl) has one chlorine atom. It is a poisonous gas. Dichloromethane (CH_2Cl_2) has two chlorine atoms. It is a colorless liquid often used as a pesticide. With three chlorine atoms, trichloromethane ($CHCl_3$) is better known as chloroform. It was one of the first anesthetics—a drug that makes people unconscious. Tetrachloromethane (CCl_4) has four chlorine atoms and no hydrogen atoms. It is used in dry cleaning.

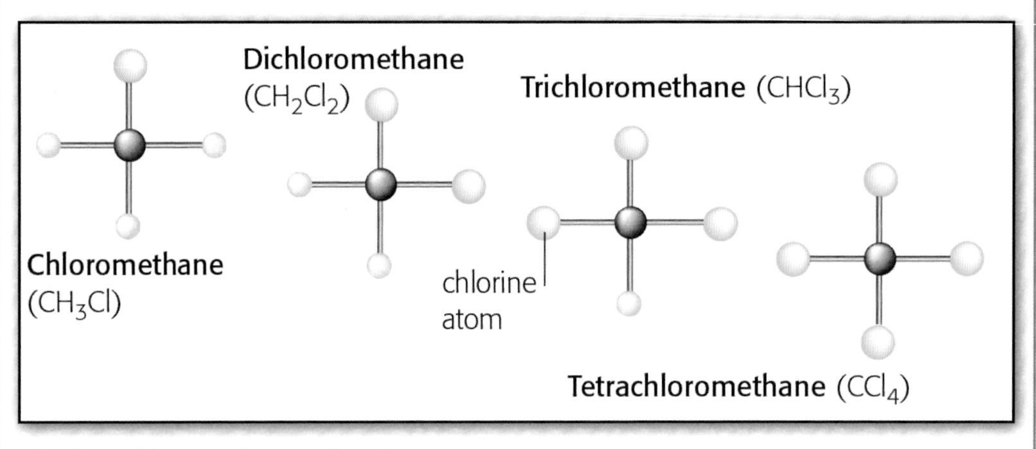

The four chloromethane molecules.

Organic halides are also very stable compounds. Fluorine is the most reactive nonmetal element of all, and the other halogens are not far behind. The bonds their atoms form with carbon atoms are very strong. As a result, the organic halides do not break down easily.

For many years, chemists thought that certain organic halides were too stable to react in normal conditions. People have since released huge amounts into the air. However, we now know that these chemicals have been very destructive to the environment.

CHEMISTRY IN ACTION

LOSING OZONE

A group of organic halides are called the chlorofluorocarbons. This name is often shortened to CFC. CFC compounds contain fluorine (F) and chlorine (Cl) atoms attached to chains of carbon atoms. CFCs are very stable compounds because the bonds between the atoms are very strong. In normal conditions, such as those in your house, they never react. If CFCs are released into the air, they will stay there for a very long time without changing.

In the 20th century, CFC gases were used in many ways. They were used inside refrigerators and to make the spray in aerosol cans. Chemists thought that CFCs could be released into the air and would not cause any problems.

However, in 1974, Mario Molina (1943–), a Mexican chemist, discovered that CFCs were reacting with ozone in the atmosphere. Ozone is an allotrope of oxygen. Most oxygen molecules have two atoms in them, making an O_2. Ozone has three atoms in an unstable O_3 molecule.

Ozone forms a layer high in the atmosphere that filters out dangerous radiation from the sun. The CFCs are destroying the ozone layer and letting harmful radiation through the atmosphere. In 1987, CFCs were banned. The ozone layer is now slowly re-forming. CFCs have been replaced with other organic compounds. For example, aerosol cans now contain the ether methoxymethane.

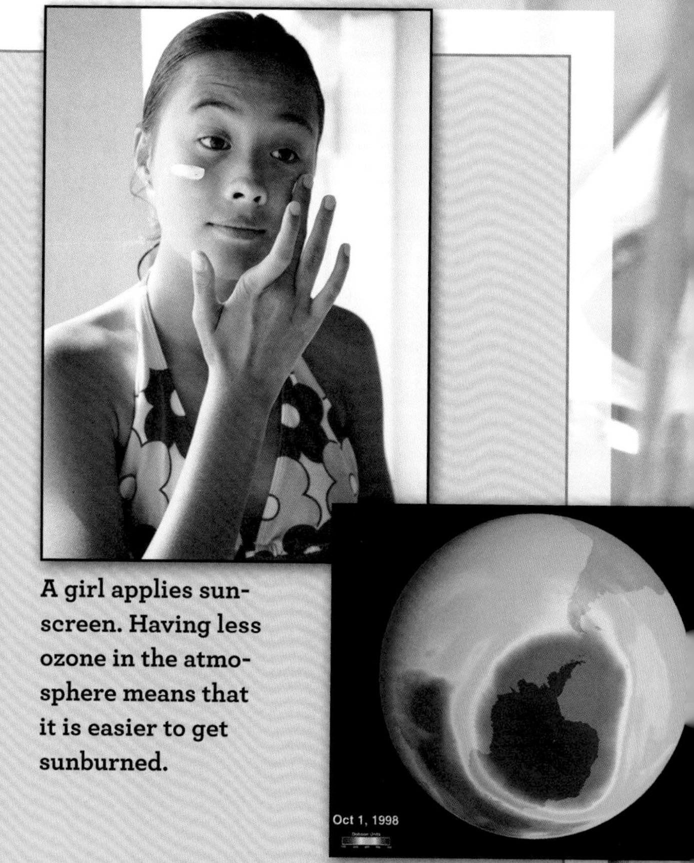

A girl applies sunscreen. Having less ozone in the atmosphere means that it is easier to get sunburned.

A satellite image of the hole in the ozone layer (purple).

Old refrigerators that contain a CFC must have the gas removed.

CHAPTER SEVEN

CHARACTERISTICS OF POLYMERS

Plastic balls are made from polymer. Polymers can be made into any shape and are used in place of all types of naturally occurring materials, such as wood, stone, glass, china, and metals.

Polymers are compounds made from long chains of smaller molecules. Polymers occur naturally in the world around us. They are also made from petrochemicals for use as plastics and to make clothing.

The world's main source of hydrocarbons is petroleum, or crude oil. Nine-tenths of the hydrocarbons in crude oil are turned into gasoline and other fuels.

What happens to the rest? Most of it is turned into compounds called polymers. Polymers are as varied as they are useful. They are used in everything from fighter planes to frying pans.

KEY DEFINITIONS

- **Compound:** A substance containing the atoms of two or more different elements.

- **Hydrocarbon:** An organic compound containing carbon and hydrogen atoms.

- **Molecule:** Two or more atoms connected together.

- **Organic:** Describes a compound that contains carbon and generally hydrogen but also atoms of other elements.

TAKING A CLOSER LOOK

MONOMERS

Polymers are chains of smaller molecules called monomers. A polymer may be made up of just one type of monomer. Such polymers are called homopolymers. (The word *homo* means "same.") Other polymers contain two or more types of monomers that are bonded alternately, one after the other. These polymers are called copolymers. (The word *co* means "together.")

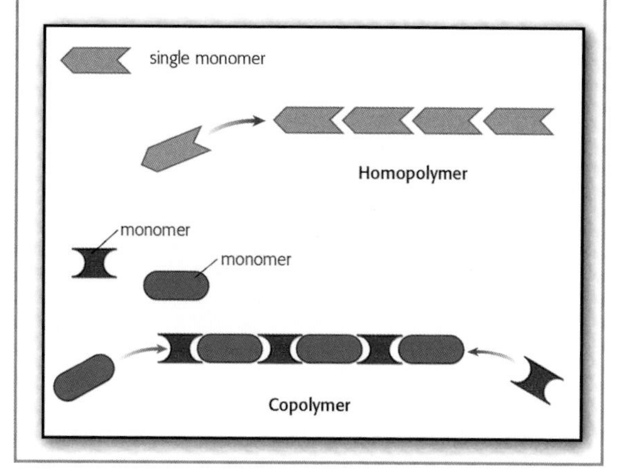

MOLECULAR CHAINS

Polymers are very large molecules. They are made up of many smaller molecules that are connected into a chain. The smaller units are called monomers. The word *mono* means "alone," while *poly* means "many." So, a polymer is a compound that contains many monomers. Monomers form polymers in a process called polymerization.

DIFFERENT KINDS OF POLYMERS

There are many types of polymers. Some occur in nature, but people make most of the polymers around us from petrochemicals. Many polymers are known as plastics. They can be molded into any shape. Other polymers are called rubbers. They are stretchy, or elastic, and can be bent out of shape easily but always spring back to their original form.

The properties of polymers are determined by the nature of their monomers and the way the chains are formed.

We use polymers every day. For example, in this shopping cart the cardboard and plastic packages are made from polymers. The food itself contains polymers, and even the cart's wheels are made from a polymer.

CREATING POLYMERS

Antique golf balls made from gutta percha, a natural plastic made from tree sap.

Monomers can be made to polymerize (form into a polymer) through a number of reactions. Monomers that have double bonds, such as alkenes, are polymerized by an addition reaction. These reactions occur when a double bond breaks and forms new single bonds.

The simplest polymer made in this way is polyethylene, known by chemists as polyethene. The monomer of this polymer is ethene (C_2H_4). Ethene has a double bond joining its two carbon atoms. During polymerization, the double bond breaks, and the carbon atoms from a long chain. Each carbon atom is bonded to two other carbon atoms and two hydrogen atoms.

CHEMISTRY IN ACTION

NATURAL POLYMERS

Nature is full of polymers. The bodies of plants, such as the wood inside trees, are made from a polymer called cellulose. This compound is a chain of sugar monomers. Starch is another polymer made from sugars. It is the soft material in bread, potatoes, and rice.

Even genes, the coded instructions that control how living bodies grow, are a polymer. That polymer is called deoxyribonucleic acid (DNA). It is made using four monomers. Each gene is coded with a unique combination of these monomers.

Rubber is also a natural polymer. It comes from latex, the sticky white sap of the rubber tree. Adding acid and salt separates the solid polymer from the liquid part of the sap. At this stage raw rubber is stringy, like the cheese on a pizza. A process called vulcanization makes the rubber much tougher.

Latex is tapped from a rubber tree. Rubber made from latex is now often replaced with similar polymers made from hydrocarbons.

CHEMISTRY IN ACTION

WHAT ARE PLASTICS?

People use most of the polymers discussed in this chapter to make plastics. Plastics are very useful materials because they can be made into any shape. The word *plastic* comes from the Greek word for "mold."

Plastics have many advantages over other materials. For example, plastics do not rust or corrode like metals. They can be made to be flexible, so they do not shatter like glass. They are also waterproof, unlike wood. Plastics are also good insulators—they do not carry an electric current. Electric cables are coated in plastic to make them safe to handle.

A few rubberlike natural polymers can be made into plastic, but plastics are really human-made materials. The first plastics were invented in the late 19th century. These early plastics were brittle (shattered easily) and expensive to make, so they were not used as widely as plastics are today. Plastics today are very inexpensive and are used in everything from spacecraft to shopping bags.

There are two types of plastics: thermoplastics and thermosets. The word *thermo* means "heat." A thermoplastic becomes softer and more moldable when it gets hot. It eventually melts and can be remolded into any shape. Polyethylene and PVC are thermoplastics.

A thermoset is the opposite of a thermoplastic. It gets harder as it is heated. Once it has hardened, a thermoset will not melt. Thermosets are molded into objects that need to stay rigid when hot. Polyester and rubber are thermosets.

MODERN POLYMERS VERSUS TRADITIONAL MATERIALS

Application	Polymer	Traditional material	Advantages of polymer
Molded objects	Polypropene	Metal	This plastic is rigid like metal but much lighter. It can also be molded at lower temperatures.
Bottles	PET (Polyethylene terephthalate)	Glass	PET is lighter than glass, and it does not shatter when dropped.
Windows	Polycarbonate	Glass	This polymer makes windows that will not shatter. However, it gets scratched more easily than glass.
Paints	Acrylic	Oil	Acrylic paints do not smell as strongly as oil paints, and they do not crack when dry.
Clothes and fabrics	Nylon	Cotton and wool	Nylon is not damaged by heat and water, and it can be woven into huge sheets.

CHEMISTRY IN ACTION

ARTIFICIAL FIBERS

Our clothes are made from fibers that are woven together. For thousands of years, people used fibers made from natural polymers. For example, wool comes from the fur of sheep and goats, while cotton fibers are made from the fluff around the seeds of cotton plants.

These natural fibers are often quite short, and they have to be spun together to make threads and yarns long enough for weaving. In the late 19th century, chemists developed ways of making stronger fibers that were made from much longer polymers.

The first artificial fiber was made from cellulose, the polymer in wood. Fabric made from cellulose is called rayon. In the 1930s, U.S. chemist Wallace Carothers (1896–1937) invented nylon. Nylon was a completely new polymer made from amines. Nylon has become the most common artificial fiber. It can be made into a huge variety of objects, from silky sheets to the bristles of a brush.

A strand of nylon is drawn from a beaker full of the polymer.

Polyethylene can be made into long, straight chains or branched networks. The straight chains produce a hard and stiff material. Objects made from branched chains are more flexible.

Wetsuits are made of neoprene. This is a waterproof rubber made by addition reactions.

POLYMER NAMES

Other polymers made by addition reactions include polypropene and polystyrene. They are named for their monomers with *poly-* added at the beginning. Other polymers have very long names almost too long to say! Instead these polymers, are known by their initials. For example, PVC stands for polyvinyl chloride. Vinyl chloride is another name for chloroethene (C_2H_3Cl).

TAKING A CLOSER LOOK
COMMON POLYMERS

You may have heard the names of some common polymers, such as PVC or polyethylene. These compounds and other polymers have a range of different properties. Many of their properties depend on the nature of their monomers. Monomers are the small units that join together to make a long chain. Many plastics are made from a mixture of polymers. Each polymer adds certain characteristics to the plastic.

Polymer	Monomer	Monomer structure	Properties of polymer
Polyethene (polyethylene)	Ethene	carbon hydrogen	Polyethene makes flexible plastics. It is used in packaging and to insulate electrical wires.
Polypropene (polypropylene)	Propene		This polymer makes plastics that are similar to polyethene but slightly tougher and more expensive.
Polystyrene	Styrene	phenyl	This polymer is used to make styrofoam. It is also added to other polymers to make them waterproof.
PVC (polyvinyl chloride)	Chloroethene (vinyl chloride)	chlorine	PVC makes very tough plastics. They are not damaged by fire or strong chemicals and are good insulators.
Teflon (polytetrafluoroethene)	Tetrafluoroethene	fluorine	Teflon is a very slippery substance that is used in nonstick pans.

COPOLYMERS

Polymerization carries on until there are no monomers left. If a new monomer is then added, the chain grows again. This behavior makes it possible to produce copolymers made from two or more different monomers.

The properties of a copolymer depend on the different monomers that make it up. Ethene makes soft polymers, while polypropene is tougher. Styrene makes glassy polymers, while rubber polymers are elastic. Chemists can blend these monomers to produce a polymer with just the correct amount of each property. Copolymers can be made from blocks of each monomer, or they can be arranged randomly. Polymers can also be produced with a precise arrangement

CHEMISTRY IN ACTION

TOO SLIPPERY TO STICK

Nonstick pans are coated in a polymer called Teflon. This name is short for polytetrafluoroethene. Teflon is the slipperiest solid known. That is why nothing sticks in a nonstick pan—all the food just slips off the coating of Teflon.

Teflon was invented by the DuPont company in 1938, the same U.S. organization that had invented nylon a few years earlier. As well as nonstick kitchenware, Teflon has many other uses. For example, spacesuits and other gear used by astronauts contain Teflon. Teflon is used to make Gore-Tex clothes. These are made from two layers of nylon with Teflon sandwiched in between. Gore-Tex clothes are waterproof in a very special way. They stop rainwater from getting in, but the wearer's sweat can pass out in the other direction.

The inside of a nonstick frying pan is coated with a layer of Teflon. The Teflon is so slippery that even burned food will not stick to it.

of monomers. For example, two types of monomers could be arranged one after the other. However, polymers like that are more expensive to produce.

Large sewer pipes made from PVC. PVC makes tough objects that do not corrode or become damaged easily.

Acrylic fiber is an example of a copolymer. It is a blend of two esters of certain acrylic acids—very reactive types of carboxylic acids.

POLYMERS AND CONDENSATION REACTIONS

Some monomers do not form polymers by addition reactions. Instead they polymerize with a condensation reaction. This reaction produces a molecule of water (H_2O) as the monomers bond together.

Nylon, polyester, and the natural polymers cellulose and starch are condensation polymers. Their monomers have two or more functional groups. The monomers join together when the functional groups on each monomer react and form bonds.

The monomers have at least one more functional group, which is not involved in holding the polymer together. These free groups can form bonds with monomers on different polymers. That

KEY DEFINITIONS

- **Functional group:** A section of an organic molecule that gives it certain chemical properties.
- **Polymerization:** The process that makes monomers join together to make polymers.

A VERSATILE MATERIAL?

You would be right to think that plastic is a very useful type of material. After all, it can be made into just about anything. But what happens when we put plastic in the garbage? One of the properties of plastic is that it does not decay easily. So plastics buried along with other garbage remain unchanged for many thousands of years.

Other materials, such as metal or wood, can be recycled or reused. Plastic is more difficult to recycle. Thermosets cannot be melted down at all, and a mixture of different thermoplastics must be separated out before they can be remolded.

Empty plastic bottles lie in a recycling plant. Waste plastic does not weigh very much, but it takes up a huge amount of space.

creates crosslinks between several polymer chains and makes a very strong network.

PROPERTIES OF POLYMERS

When you look at a plastic cup, a rubber ball, or a nylon rope, you cannot see the polymers inside them. The polymers are obviously far too small to see. If you could see them, they would not all look the same. For example, a plastic cup is very different from a rubber ball. That is because their polymer chains are arranged differently.

The properties of an object made from a polymer depend on how the polymers are arranged. The simplest arrangement is to have polymers that are unbranched and form straight chains. The chains may be many thousands or even millions of atoms in length. A sample of these polymers will contain chains with a wide range of lengths.

Unbranched, straight polymers are packed close together. Some even form crystals. Polymers like that make stiff materials. They do not change shape easily because the polymers inside are packed too closely to move around much.

However, when a sample of this material is stretched, the polymers slip past one another. As a result, the sample becomes longer. When the stretching stops, the polymers stay in their new position, and the sample keeps its new stretched shape. Materials that behave like this are said to be plastic.

Styrofoam is polystyrene pumped full of air bubbles.

A branched polymer has side chains along its main chain. The chains stop polymers from packing together closely. As a result, the polymers are more flexible. Addition polymers, such as polyethylene, can be formed as either straight or branched chains.

CROSSLINKS AND COILED CHAINS

Rubbers are polymers that have coiled chains. When they are stretched, the coils straighten and become longer. However, once the stretching ends, the coiled chains spring back to their original shape. Polymers that behave like that are described as being elastic.

CHEMISTRY IN ACTION

VULCANIZATION

When it is first made, rubber is a stretchy gumlike material. To make it into a useful elastic material, which can be used to make tires, soles for shoes, and countless other things, the rubber must be vulcanized. Vulcanizing adds crosslinks between the rubber's polymers. In the most common vulcanization process, sulfur is added to the rubber and heated. A sulfur (S) atom bonds to a carbon (C) atom on two polymers. That creates a C—S—C crosslink between chains. Adding more crosslinks makes the rubber even tougher.

Tires made from vulcanized rubber are very tough, and that makes them difficult to dispose of. In the state of New York alone, 18 million used tires are thrown away each year.

TAKING A CLOSER LOOK

PLASTIC OR ELASTIC?

The stretchiness of plastics, rubber, and other materials made from polymers depends on how their polymers are arranged.

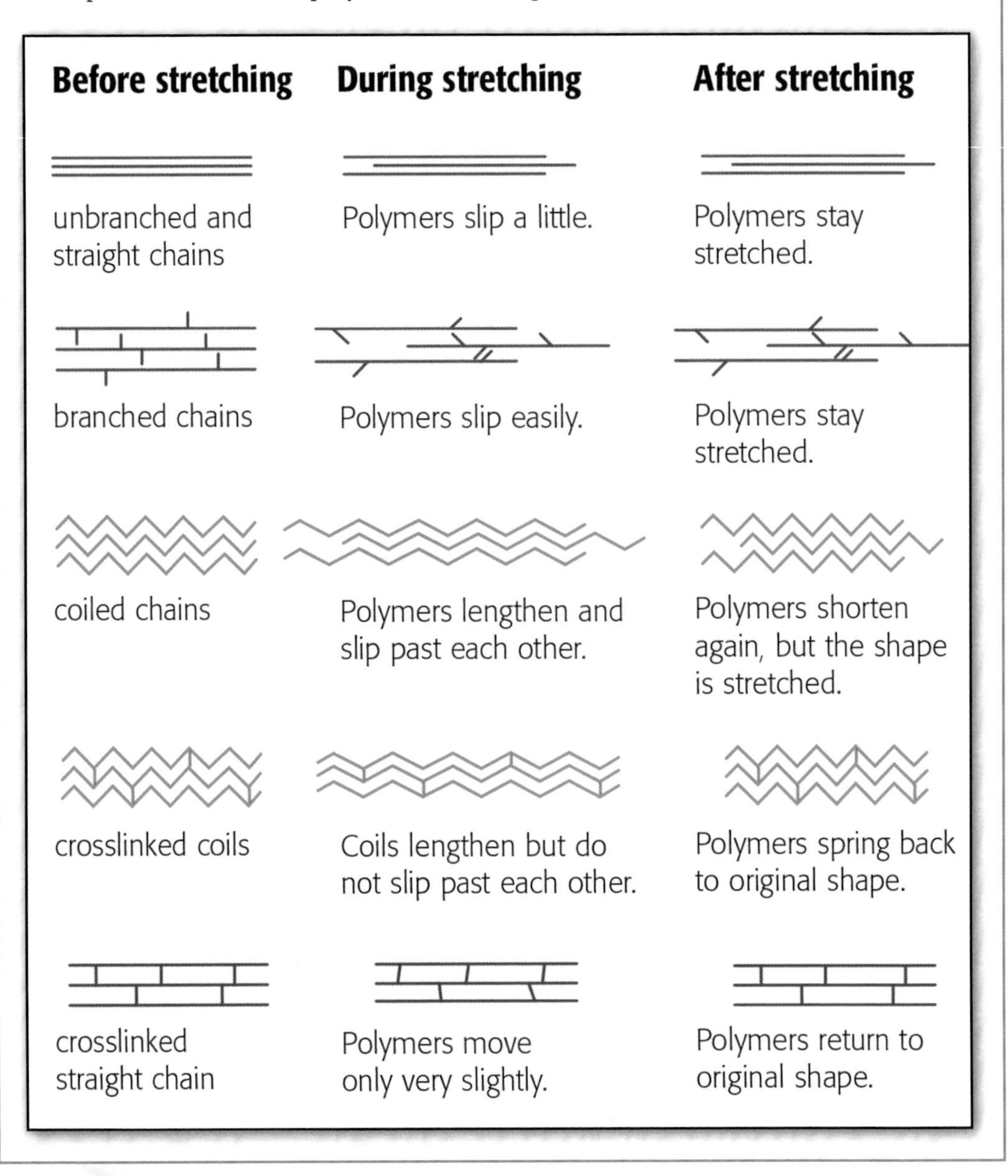

KEY DEFINITIONS

- **Bond:** An attraction between atoms.
- **Crosslink:** A bond between two polymers.
- **Crystal:** A solid made of regular repeating patterns of atoms.
- **Elastic:** Describes substances that return to their original shape after being stretched.
- **Plastic:** Describes substances that change shape permanently after being stretched.

An antique radio made of Bakelite was one of the first thermosets to be invented.

However, untreated rubber behaves like a plastic as well. Some of the polymers slip past each other, causing the rubber to stretch permanently. Adding crosslinks between the coils stops that from happening and makes the rubber completely elastic. The process that adds crosslinks to rubber is called vulcanization.

HEATING POLYMERS

Polymers with lots of crosslinks make very rigid materials. For the material to break or change shape, the bonds in the crosslinks must be broken. Many polymers form crosslinks when they are heated. These compounds are called thermosets.

Thermosets are used to make molded objects. The powdered ingredients of the polymer are packed into a mold and heated. The heat makes the thermoset polymers form. The heat also causes crosslinks to connect the polymers making the object very rigid.

CHAPTER EIGHT

BIOGRAPHY: FRIEDRICH WÖHLER

As mentioned previously, one of the fathers of organic chemistry was German chemist Friedrich Wöhler. By accidentally synthesizing urea, a chemical everyone thought could only be produced by living things, he proved that chemistry was chemistry, whether it happened inside living creatures or in an

Frankfurt's Römerberg Square and Fountain of Justice are seen here as they appear today. Wöhler's family moved to Frankfurt in 1812.

ON THE RUN?

According to the biographical sketch of Wöhler written by Professor Charles A. Joy for the *Popular Science Monthly* in 1880, the Wöhler family was in Escherheim because they'd had to beat a hasty retreat from Hesse-Cassel, once an electorate, or state, of the Holy Roman Empire. Friedrich's father, Auguste, had formerly been Master of Horse for an important nobleman, the Elector of Hesse-Cassel, who, Joy writes, "is celebrated in history for the violence of his temper."

As Joy tells it, one day Auguste got so fed up with his employer's insulting behavior that he took out a horsewhip and beat the Elector with it until he ran off. Then Auguste jumped on the fastest horse in the stables and made his escape, accompanied by a groom who would later return the horse.

"The august Elector," Joy notes, "feeling ridicule, thought it wisest to let the matter drop, and never pursued the fugitive."

inanimate object. But that was just *one* of the important discoveries by this amazing scientist.

Friedrich Wöhler was born on July 31, 1800, in the hamlet of Escherheim (now a suburb of Frankfurt, Germany), in the house of his uncle, the village pastor. Shortly after Friedrich's birth his father bought an estate in Rodelsheim, where the family lived until 1812; after that, they lived in Frankfurt.

Friedrich's father, Auguste, was a well-educated man, and his mother, Katherina Schröder, was the daughter of a professor of philosophy, so Friedrich received a first-rate education. He learned writing, drawing, and French from his father even before starting school at age seven, and also had private lessons in Latin and music.

Very early on he showed an interest in conducting experiments and making collections, especially of interesting minerals. A friend of his father's who had studied science at the University of Göttingen had quite a bit of scientific equipment that he showed the boy how to use.

TEENAGE YEARS

At age 14, Friedrich began attending the Frankfurt Gymnasium (high school). Its focus was on a classical education, and Friedrich, as Professor Joy puts it, "was not distinguished for much knowledge of the ancient languages." Fortunately, though, one of the instructors, Dr. Buch, also had an interest in chemistry, physics, and mineralogy, and had set up an improvised laboratory in his kitchen where students were occasionally allowed to conduct experiments.

Working with Friedrich, Buch analyzed some pyrites from Bohemia, finding they contained the recently discovered element selenium. He published the results of his analysis, and shared credit with Friedrich, who thus had his name on his first scientific paper while still a teenager.

Friedrich worked with Dr. Buch in examining pyrites, such as the sample of natural crystal pyrite seen here, and they determined that they contained selenium. Buch acknowledged Friedrich's contribution as co-author on the published scientific report.

By age 18, Friedrich's interest in chemistry had grown into a passion. Joy writes that "his room at home was transformed into a laboratory full of glasses, retorts (a glass container with a long neck that is used in chemical processes), washing-bottles, and minerals—everything in the greatest confusion."

His schoolboy friend Hermann von Meyer owned a chemistry textbook written by Ferdinand Wurzer that Friedrich knew almost by heart, and it was to von Meyer that Friedrich wrote letters detailing many of his experiments, including oxygen preparation, phosphorus extraction, and the isolation of potassium using a powerful battery. (He and his family may have been lucky to survive: over the course of his experiments, he suffered severe phosphorus burns and at one point almost suffocated after he broke a flask full of poisonous chlorine.)

Friedrich didn't spend *all* his time in his makeshift laboratory, though: his father, who thought his son had a rather weak constitution, made sure he got plenty of exercise, including riding, fencing, swimming and boating.

Wöhler graduated from the Gymnasium at Easter, 1820, and enrolled in a medical degree course at Marburg University, where Wurzer, author of the chemistry textbook that had belonged to his friend von Meyer, was professor of medicine. Wöhler studied botany, zoology, mineralogy, physics, and anatomy, and began to perform dissections, but his passion remained chemistry. He even transformed his parlor into a laboratory (much to the annoyance of his landlord).

One day he discovered iodide of cyanogen. (Unfortunately it had already been discovered, but he didn't know that.) Very excited, he told Wurzer about it. Instead of praising him, Wurzer reprimanded him for neglecting his medical studies in favor of his own experiments. That reprimand may have had something to do with the fact that the next year Wöhler transferred to Heidelberg University.

There, though he was still studying medicine, he obviously impressed the chemistry professor, Leopold Gmelin, who told him not to bother coming to lectures but rather to assist in Gmelin's laboratory.

A portrait of Friedrich Wöhler shows him as a young chemistry professor during the late 1820s in Berlin.

Although Wöhler graduated with a degree in medicine in 1823, Gmelin discouraged him from becoming a doctor. Instead, he urged Wöhler to devote himself to chemistry, and helped him get a year-long job in Stockholm as an assistant to the famed Swedish chemist Jöns Jacob Berzelius, one of the fathers of modern chemistry. (Among other achievements, he worked out the modern method of chemical notation.)

Wöhler left Sweden in September 1824. He lived in Heidelberg for a while, then in March 1825, with a little help from Berzelius, he became chair of chemistry at the brand-new Gewerbeschule (a vocational school) in Berlin. He made the first of his significant discoveries there in 1827, becoming the first (possibly) to isolate aluminum.

But 1828 proved to be an even more momentous year. Not only did he become a professor and marry (his cousin Franziska Wöhler, the daughter of a wealthy banker from the city of Cassel), but also he made the most important discovery of his career—in fact, one of the most important discoveries of the nineteenth century.

VITALISM

For centuries, a philosophy called vitalism had held sway in learned circles. Vitalists believed that living organisms are fundamentally different from nonliving things: that they contain a nonphysical element, a "vital spirit" that gives them life and, in turn, means they are governed by distinct scientific principles from nonliving things.

But during the 16th and 17th centuries, as modern science began to emerge, some scientists rejected this idea. The French philosopher and mathematician René Descartes said he thought that animals—and the human body—are essentially nothing more than "automata," complex mechanical devices.

Others, however, thought that Descartes's mechanistic view couldn't explain movement, perception, development, or life. Among those were Wöhler's mentor, Berzelius. At one point he wrote that no one would ever be able to synthesize the products of living organisms in the laboratory.

Wöhler, however, did just that. Not that that was what he set out to do. He was working with cyanic acid and ammonia, attempting to create a new compound, ammonium cynate. But the result was always the same: oxalic acid...and a strange material he couldn't identify, a white substance made up of "colorless, clear crystals often more than an inch long in the form of slender, four-sided dull-pointed prisms."

Trying to figure what the strange stuff was, he discovered that it reacted to various chemical processes the same way as urea, a chemical compound found in urine. According to the vitalists, urea could only be produced by living things because

A DAY IN THE LAB

Wöhler continued to write lots of letters, one of which, quoted by A.W. Kahlbaum in a 1900 biography, gives a glimpse of a typical day in Berzelius's laboratory:

"There (Berzelius) sits at his table, which is covered with glass and platinum—his present work is on fluorides that can only be done in platinum. He wears a black coat—on which several reactions can be studied...I sit down at the table, look at my lithium salts crystallising from yesterday, and get on with an analysis of a cyanic salt compound which I had the misfortune to discover and must now follow up. I carry on with the analysis of lievrite which I have carried out six times already but still have not finished because the results never agree..."

This German stamp commemorates the one hundredth anniversary of Wöhler's death and honors his synthesis of urea. His experiment was a landmark discovery by proving that one could synthesize an organic compound by using inorganic substances.

it was clearly a product of the supposed "vital force." But through further tests, Wöhler confirmed that the substance he had produced was, in fact, urea.

"This unexpected result," he wrote (as quoted by Professor T.E. Thorpe in an essay originally published in 1902) "is a remarkable fact, in so far as it presents an example of the artificial formation of an organic body, and indeed one of animal origin, out of inorganic materials."

To Berzelius, Wöhler put the result in much plainer language. Robert Carlson, in his book *Biology Is Technology*, quotes him this way: "I can prepare urea without requiring a kidney of an animal either man or dog." He went on to write that he had witnessed "the great tragedy of science, the slaying of a beautiful hypothesis by an ugly fact."

The "beautiful hypothesis" was vitalism. Professor August W. von Hofmann, an early biographer of Wöhler, wrote that "the synthesis of urea was an epoch-making discovery in the real sense of the word. With it was opened out a new domain of investigation, upon which the chemist instantly seized."

A DAY IN THE LAB

Wöhler's experiment wasn't immediately hailed as a nail in the coffin of vitalism: actually, at the time, chemists were more excited about the fact that it reinforced the startling new concept of isomerism.

Isomerism had been the great discovery of Justus von Liebig, a young professor (three years younger than Wöhler) at the University of Geissen who would soon become a great friend and colleague of Wöhler's. Liebig had shown that fulminic acid, a powerful explosive, and cyanic acid, a harmless compound, had the same chemical composition despite their radically different properties. The idea that compounds could contain the same elements in the same proportions and yet exhibit completely different properties was a huge step on the road to modern chemistry.

The term "isomerism" was coined by Berzelius, who at first scoffed at the notion but was eventually forced to accept it.

Professor Justus von Liebig's laboratory at the University of Giessen was one of the first practical teaching laboratories in the world where many scientists did their research and experiments.

That "new domain" was what people today call organic chemistry.

MORE DISCOVERIES TO COME

At age 28, Wöhler had already ensured his place in the history books, and he was far from finished. Between 1821 and 1880, more than 300 papers appeared bearing his name. (He probably could have attached his name to many more, but he didn't believe in adding his name to students' work carried out under his supervision.)

In 1831, Wöhler resigned his professorship and moved with his wife and young son and daughter to Cassel, where his wife's family lived. He took up a position similar to the one he'd had in Berlin at a new Gewerbeschule.

He probably made the move because of his wife's ill health: she died in 1832. Within a few months he was married again, this time to Julie Pfeiffer, daughter of a wealthy banker.

The same year he moved to Cassel, he and Liebig published another of the most famous papers in the history of

The town hall of Göttingen is pictured here in this print from the 19th century. Wöhler became a chemistry professor at the University of Göttingen in 1836.

chemistry, on the oil of bitter almonds (benzaldehyde). The two chemists determined that a long series of different compounds that could be produced from the oil all contained a single unchanging constituent, which they called "benzoyl." This discovery introduced the concept of radicals: chemical compounds that act as if they are elements.

In early 1836, Wöhler was appointed professor of chemistry at the University of Göttingen. He continued his research both on his own and with Liebig. In their last great work together, the two friends discovered more than 15 new chemical compounds while investigating uric acid.

The work related closely to Wöhler's original synthesis of urea, and showed, Thorpe quotes them as writing, that it might "be regarded as not only probable but as certain" that ultimately *any* organic chemical could be produced in the laboratory.

This monument to Friedrich Wöhler is located in Göttingen and was created in 1890 by Ferdinand Hartzer.

"GOOD-HUMORED AND WELL-LOVED"

In later years, Wöhler turned most of his attention to inorganic chemistry, where he continued to make discoveries. He remained at the University of Göttingen until his retirement at age 80, and received countless awards and honors from scientific societies around the world. It's estimated that around 8,000 students passed through his classes, many of whom went on to distinguished careers in their own right.

Wöhler had a son and daughter with his first wife, and four daughters with his second wife. He died at age 82 on September 23, 1882.

As Robin Keen put it in a 1985 article for *Platinum Metals Review*, "Unlike many of the major scientists of the nineteenth century, Wöhler was a modest, unpretentious, good-humoured and well-loved man, who genuinely enjoyed teaching the students who flocked to his laboratory in Göttingen from many parts of the world."

July 31, 1800 Friedrich Wöhler is born in the hamlet of Escherheim to parents Auguste and Katherina.

1807 Wöhler, age seven, begins public school.

1812 Wöhler family moves from Escherheim to Frankfurt.

1814 Wöhler enrolls in Frankfurt Gymnasium.

1814–1820 He attends Frankfurt Gymnasium; Wöhler develops an interest in chemistry; his first scientific paper is published in conjunction with his instructor, Dr. Buch.

1820 He enrolls in a medical degree course at Marburg University.

1821 Wöhler transfers to Heidelberg University and begins working in the laboratory of Professor Leopold Gmelin.

1823 Wöhler graduates with a degree in medicine from Heidelberg University; with Gmelin's help, he obtains a one-year position in the laboratory of famed Swedish chemist Jöns Jacob Berzelius.

September 1824 Wöhler returns to Heidelberg.

March 1825 He is named chair of chemistry at the Berlin Gewerbeschule (vocational school).

1827 Wöhler is possibly the first to discover aluminum.

1828 He is promoted to professor of chemistry at the Berlin Gewerbeschule; Wöhler marries his cousin Franziska Wöhler.

1828 He synthesizes an organic chemical, urea, out of inorganic chemicals, helping to launch the field of organic chemistry; about the same time, he meets Justus von Liebig, professor of chemistry at the University of Geissen.

1831 Wöhler resigns his professorship and moves with his young son and daughter to Cassel, home to his wife's family; he takes on a similar position at the Gewerbeschule in Cassel.

1832 Franziska Wöhler dies; Wöhler marries Julie Pfeiffer, the daughter of a wealthy banker.

1832 Wöhler and Liebig publish their famous paper on the oil of bitter almonds, demonstrating that a series of compounds that could be produced from the oil all contained a single unchanging "radical," a compound that acted like an element.

1836 Wöhler is appointed professor of chemistry at the University of Göttingen, and he remains there for the rest of his career.

1880 Wöhler retires from active teaching.

September 23, 1882 Wöhler dies in Göttingen; he is survived by his second wife, a son, and five daughters.

PERIODIC TABLE OF ELEMENTS

The periodic table organizes all the chemical elements into a simple chart according to the physical and chemical properties of their atoms. The elements are arranged by atomic number from 1 to 118. The atomic number is based on the number of protons in the nucleus of the atom. The atomic mass is the combined mass of protons and neutrons in the nucleus. Each element has a chemical symbol that is an abbreviation of its name. In some cases, such as potassium,

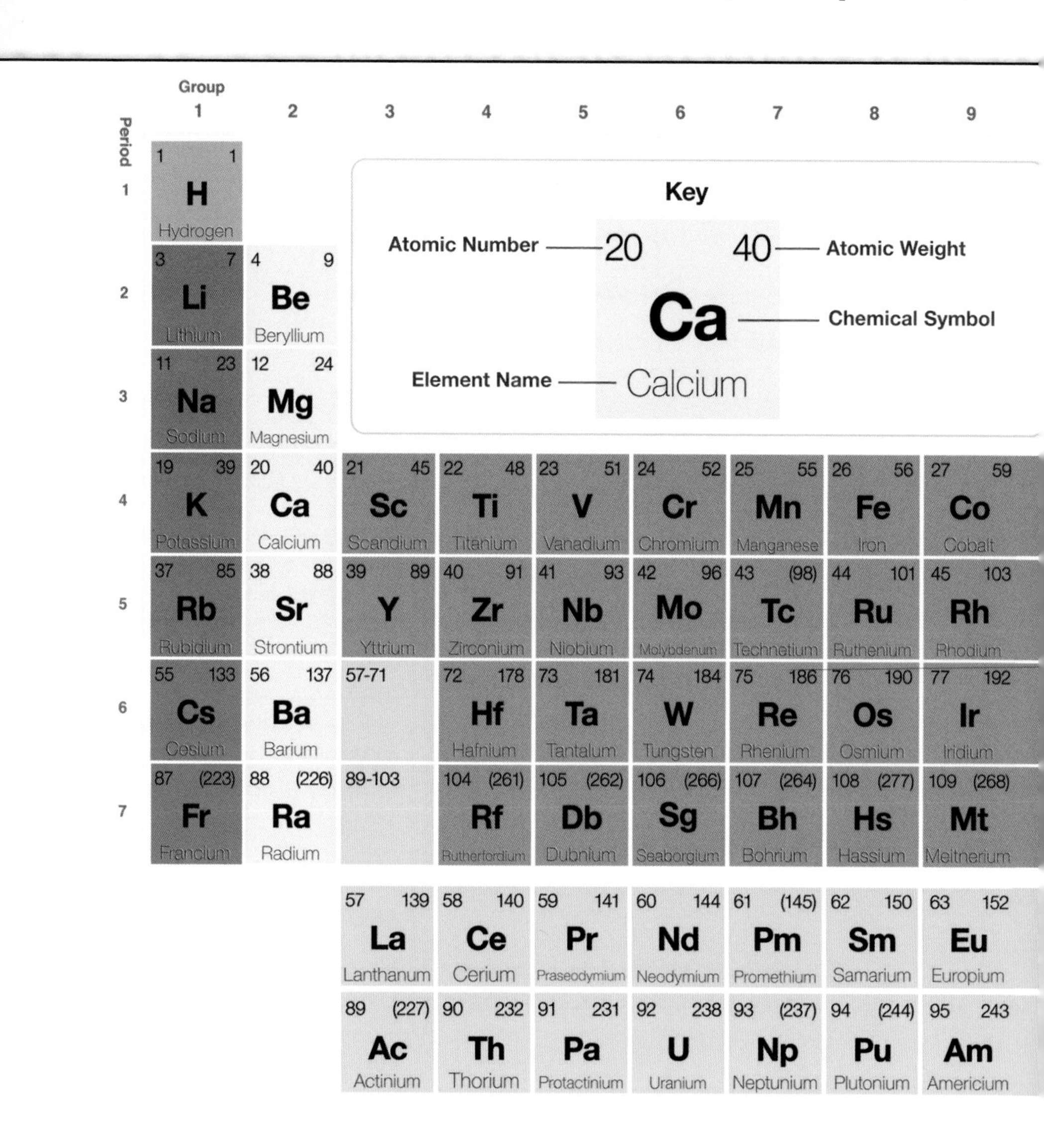

the symbol is an abbreviation of its Latin name ("K" stands for kalium). The name by which the element is commonly known is given in full underneath the symbol. The last item in the element box is the atomic mass. This is the average mass of an atom of the element.

Scientists have arranged the elements into vertical columns called groups and horizontal rows called periods. Elements in any one group all have the same number of electrons in their outer shell and have similar chemical properties. Periods represent the increasing number of electrons it takes to fill the inner and outer shells and become stable. When all the spaces have been filled (Group 18 atoms have all their shells filled), the next period begins.

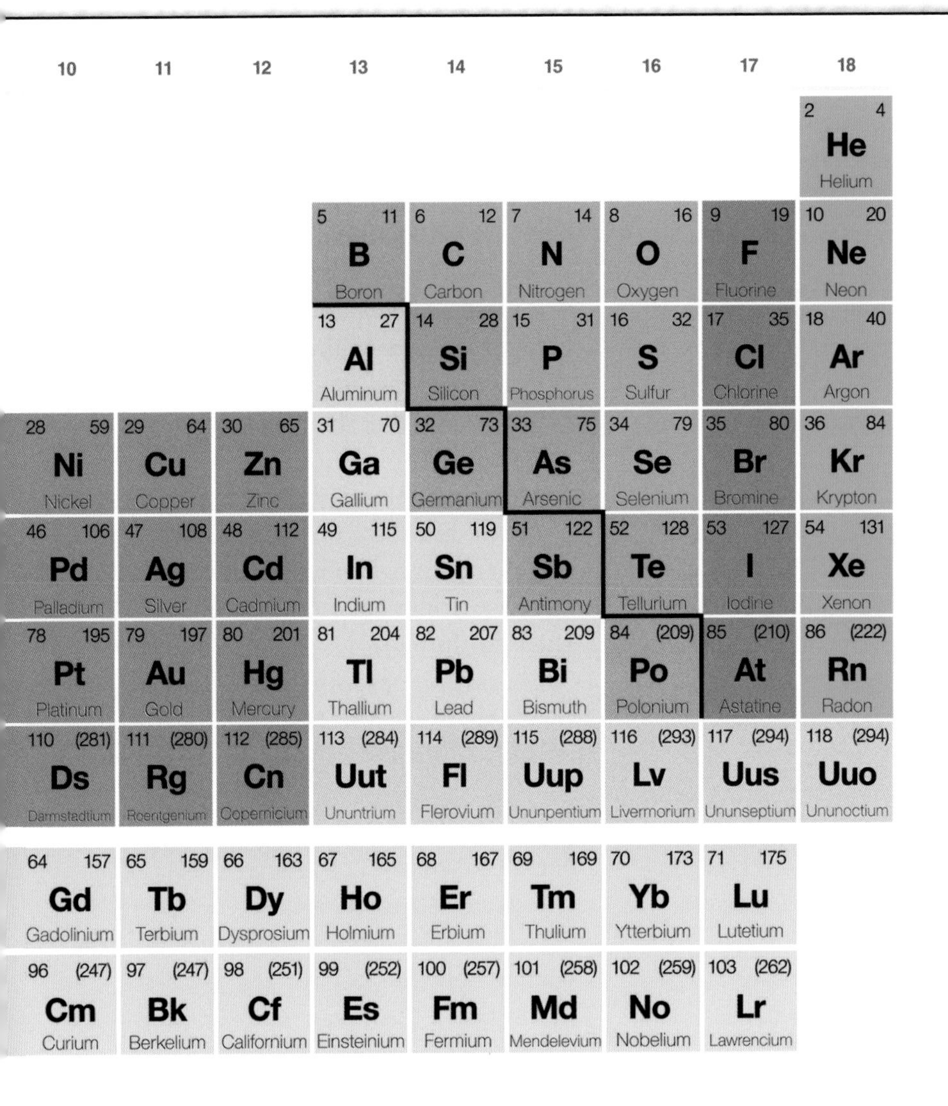

10	11	12	13	14	15	16	17	18
								2 4 He Helium
			5 11 B Boron	6 12 C Carbon	7 14 N Nitrogen	8 16 O Oxygen	9 19 F Fluorine	10 20 Ne Neon
			13 27 Al Aluminum	14 28 Si Silicon	15 31 P Phosphorus	16 32 S Sulfur	17 35 Cl Chlorine	18 40 Ar Argon
28 59 Ni Nickel	29 64 Cu Copper	30 65 Zn Zinc	31 70 Ga Gallium	32 73 Ge Germanium	33 75 As Arsenic	34 79 Se Selenium	35 80 Br Bromine	36 84 Kr Krypton
46 106 Pd Palladium	47 108 Ag Silver	48 112 Cd Cadmium	49 115 In Indium	50 119 Sn Tin	51 122 Sb Antimony	52 128 Te Tellurium	53 127 I Iodine	54 131 Xe Xenon
78 195 Pt Platinum	79 197 Au Gold	80 201 Hg Mercury	81 204 Tl Thallium	82 207 Pb Lead	83 209 Bi Bismuth	84 (209) Po Polonium	85 (210) At Astatine	86 (222) Rn Radon
110 (281) Ds Darmstadtium	111 (280) Rg Roentgenium	112 (285) Cn Copernicium	113 (284) Uut Ununtrium	114 (289) Fl Flerovium	115 (288) Uup Ununpentium	116 (293) Lv Livermorium	117 (294) Uus Ununseptium	118 (294) Uuo Ununoctium
64 157 Gd Gadolinium	65 159 Tb Terbium	66 163 Dy Dysprosium	67 165 Ho Holmium	68 167 Er Erbium	69 169 Tm Thulium	70 173 Yb Ytterbium	71 175 Lu Lutetium	
96 (247) Cm Curium	97 (247) Bk Berkelium	98 (251) Cf Californium	99 (252) Es Einsteinium	100 (257) Fm Fermium	101 (258) Md Mendelevium	102 (259) No Nobelium	103 (262) Lr Lawrencium	

acid A substance that dissolves in water to form hydrogen ions (H+). Acids are neutralized by alkalis and have a pH below 7.

alcohol A compound formed when a hydroxide ion (OH–) replaces a hydrogen atom on a hydrocarbon chain or an aromatic ring.

aldehyde A compound with a carbonyl group attached to the end of its molecule.

aliphatic compound An organic compound that has an open-chain structure, such as an alkane.

alkali A substance that dissolves in water to form hydroxide ions (OH-). Alkalis have a pH greater than 7 and will react with acids to form salts.

alkane A hydrocarbon chain in which all atoms are connected by single bonds.

alkene A hydrocarbon chain in which at least two carbon atoms are connected by a double bond.

alkyne A hydrocarbon chain in which two carbon atoms are joined by a triple bond.

allotrope A different form of an element in which the atoms are arranged in a different structure.

arene A type of hydrocarbon compound that has ringed molecules.

aromatic Describes a compound that contains one or more benzene rings.

atom The smallest independent building block of matter. All substances are made of atoms.

atomic mass number The number of protons and neutrons in an atom's nucleus.

atomic number The number of protons in a nucleus.

benzene A ring of carbon atoms in which electrons are shared by all atoms in the molecule.

bond A chemical connection between atoms.

by-product A substance that is produced when another material is made.

carbohydrate One of a group of compounds that includes sugars, starch, and cellulose. Some are essential in turning food to energy. Others are energy stores in plants, and still more build plant and animal cell membranes.

carbonyl A functional group in which an oxygen atom is connected to a carbon atom by a double bond.

catalyst Substance that speeds up a chemical reaction but is left unchanged at the end of the reaction.

chemical equation Symbols and numbers that show how reactants change into products during a chemical reaction.

chemical formula The letters and numbers that represent a chemical compound, such as "H_2O" for water.

chemical reaction The reaction of two or more chemicals (the reactants) to form new chemicals (the products).

chemical symbol The letters that represent a chemical, such as "Cl" for chlorine or "Na" for sodium.

combustion The reaction that causes burning. Combustion is generally a reaction with oxygen in the air.

compound A substance made from more than one element and that has undergone a chemical reaction.

copolymer A polymer made from two or more different types of polymer.

covalent bond A bond in which atoms share one or more electrons.

cracking A process by which products of fractional oil distillation are broken down into simpler hydrocarbons.

crosslink A bond between two polymers.

dissolve To form a solution.

electron A tiny, negatively charged particle that moves around the nucleus of an atom.

element A material that cannot be broken up into simpler ingredients. Elements contain only one type of atom.

energy level Electron shells represent different energy levels. Those closest to the nucleus have the lowest energy.

ester A compound formed when an alcohol reacts with a carboxylic acid.

ether A compound in which two hydrocarbon molecules are connected by a single oxygen atom.

evaporation The change of state from a liquid to a gas when the liquid is at a temperature below its boiling point.

fractional distillation The process of heating crude oil to separate different hydrocarbon components.

fullerenes Ball- or tube-shaped allotropes of carbon made of hexagonal or pentagonal rings of carbon atoms.

functional group A section of an organic molecule that gives it certain chemical properties.

gas The state in which particles are not joined and are free to move in any direction.

halide A compound containing a halogen atom, such as chlorine or iodine.

homopolymer A polymer made from one type of monomer.

hydrocarbon An organic compound that contains only carbon and hydrogen.

hydrogen bond A weak dipole attraction that always involves a hydrogen atom.

hydroxyl A functional group (–OH) made up of an oxygen and a hydrogen atom.

inorganic A compound that is not organic.

intermolecular bonds A bond that holds molecules together. These bonds are weaker than those between atoms in a molecule.

intramolecular bond A strong bond between atoms in a molecule.

ion An atom that has lost or gained one or more electrons.

ionic bond A bond in which one atom gives one or more electrons to another atom.

ionization The formation of ions by adding or removing electrons from atoms.

isomer A substance with the same chemical formula as another compound but which has a different structural arrangement of its atoms. It may also react differently.

ketone A compound with a carbonyl group attached in the middle of its molecule.

liquid A substance in which particles are loosely bonded and are able to move freely around each other.

matter Anything that can be weighed.

melting point The temperature at which a solid changes into a liquid. When a liquid changes into a solid, this same temperature is called the freezing point.

mixture Matter made from different types of substances that are not physically or chemically bonded together.

mole The amount of any substance that contains the same number of atoms as in 12 grams of carbon-12 atoms. This number is 6.022×10^{23}.

molecule Two or more joined atoms that have a unique shape and size.

monomer Monomers are molecules that join to form long chains called polymers.

neutron One of the particles that make up the nucleus of an atom. Neutrons do not have any electric charge.

nucleus The central part of an atom. The nucleus contains protons and neutrons. the exception is hydrogen, which contains only one proton.

organic A compound that is made of carbon and hydrogen.

organic acids The family of organic compounds that include the carboxylic acids and fatty acids, which both contain a –COOH group. Like inorganic acids, they produce hydrogen ions (H^+).

petrochemical An organic chemical that is made from petroleum or natural gas.

phenol An alcohol formed by the attachment of an –OH group to an aromatic ring.

plastic An organic polymer that can be molded or shaped into objects or films by heat.

polymerization The process that makes monomers join together to form polymers.

pressure The force produced by pressing on something.

product The new substance or substances created by a chemical reaction.

proton A positively charged particle found in an atom's nucleus.

reactant The ingredient necessary for a chemical reaction.

relative atomic A measure of the mass of an atom compared with the mass of another atom. The values used are the same as those for atomic mass.

relative molecular mass The sum of all the atomic masses of the atoms in a molecule.

salt A compound made from positive and negative ions that forms when an alkali reacts with an acid.

saturated hydrocarbon A hydrocarbon molecule in which all the carbon atoms are bonded to four other atoms. An unsaturated molecule contains double or triple bonds between carbon atoms.

shell The orbit of an electron. Each shell can contain a specific number of electrons and no more.

solid The state of matter in which particles are held in a rigid arrangement.

solute The substance that dissolves in a solvent.

solution A mixture of two or more elements or compounds in a single phase (solid, liquid, or gas).

solvent The liquid that solutes dissolve in.

specific heat capacity The amount of heat required to change the temperature of a specified amount of a substance by 1ºC (1.8ºF).

state The form that matter takes—either a solid, a liquid, or a gas.

subatomic particle A particle that is smaller than an atom.

temperature A measure of how fast molecules are moving.

American Chemical Society (ACS)
1155 Sixteenth Street NW
Washington, DC 20036
(800) 227-5558
Web site: http://portal.acs.org/portal/acs/corg/content
The ACS represents professionals in all fields of chemistry. Its Division of Organic Chemistry (http://www.organicdivision.org) specifically promotes the advancement of the field of organic chemistry.

Chemical Heritage Foundation
315 Chestnut Street
Philadelphia, PA 19106
(215) 925-2222
Web site: http://chemheritage.org/museum/index.aspx
The Chemical Heritage Foundation's museum features a permanent exhibition called "Making Modernity," which explains how chemistry has influenced people, and changing exhibits on a variety of subjects.

Chemical Institute of Canada
130 Slater Street, Suite 550
Ottawa, ON K1P 6E2
Canada
(613) 232-6252
Web site: http://www.cheminst.ca
This organization of chemists, chemical engineers, and chemical technologists works to advance the sciences in a range of areas, including health care, pharmaceuticals, energy, food, and water.

International Union of Pure and Applied Chemistry (IUPAC)
P.O. Box 13757
Research Triangle Park, NC 27709-3757
(919) 485-8700
Web site: http://www.iupac.org
The IUPAC advances the chemical sciences by fostering worldwide communication and education in the chemical sciences. It is the world authority on chemical nomenclature and standardized measurement methods.

National Academy of Sciences
500 Fifth Street NW
Washington, DC 20001
(202) 334-2000
Web site: http://www.nasonline.org
This society provides objective guidance to people in the fields of science and technology. Its Web site offers the latest news on current academic research and government policy.

National Science Foundation (NSF)
4201 Wilson Boulevard
Arlington, VA 22230
(703) 292-5111
Web site: http://www.nsf.gov
This U.S. agency promotes education and research in the sciences and engineering. Its Web site includes updates on chemistry and materials discoveries.

Youth Science Canada
1550 Kingston Road, Suite 213
Pickering, ON L1V 1C3

Canada
(866) 341-0040
Web site: http://www.youthscience.ca
This organization encourages young people to participate in project-based science by making programs available to them. Youth Science Canada holds the Canada-Wide Science Fair every year to celebrate young scientists and their experiments.

WEB SITES

Due to the changing nature of Internet links, Rosen Publishing has developed an online list of Web sites related to the subject of this book. This site is updated regularly. Please use this link to access the list:

http://www.rosenlinks.com/CORE/Orgo

Atkins, P. W. *The Periodic Kingdom: A Journey into the Land of Chemical Elements*. Reprint ed. New York, NY: Barnes & Noble, 2007.

Bateman, Graham, ed. *Organic Chemistry and Biochemistry* (Facts at Your Finger Tips). New York, NY: Brown Bear Books, 2011.

Belval, Brian. *The Carbon Elements* (Understanding the Elements of the Periodic Table). New York, NY: Rosen Publishing, 2010.

Bettelheim, Frederick A., and William H. Brown, Mary K. Campbell, and Shawn O. Farrell. *Introduction to General, Organic, and Biochemistry*. 9th ed. Independence, KY: Cengage Learning, 2010.

Burgess, Mark. "The Birth of Biochemistry." *Biochemist*, April 2011, pp. 70–71. Retrieved June 27, 2013 (http://www.biochemist.org/bio/03302/0070/033020070.pdf).

Carlson, Robert. *Biology Is Technology: The Promise, Peril, and New Business of Engineering Life*. Cambridge, MA: Harvard University Press, 2010. Chapter 4. Retrieved June 27, 2013 (http://www.biologyistechnology.com/chapter-4.html).

Cotton, Simon. *Every Molecule Tells a Story*. Boca Raton, FL: CRC Press, 2012.

Curran, Greg. *Homework Helpers: Chemistry*. Pompton Plains, NJ: Career Press, 2011.

Freinkel, Susan. *Plastic: A Toxic Love Story*. New York, NY: Houghton Mifflin Harcourt Publishing Company, 2011.

"Freidrich Wöhler." *Scientific American Supplement*, Vol. XIV, No. 362, December 9, 1882. Project Gutenberg. Retrieved June 27, 2013 (http://www.gutenberg.org/files/8687/8687-h/8687-h.htm#7).

Gregersen, Erik, ed. *The Britannica Guide to the Atom* (Physics Explained). New York, NY: Britannica Educational Publishing and Rosen Educational Services, 2011.

Guch, Ian, and Kjirsten Wayman. *Complete Idiot's Guide to Organic Chemistry*. New York, NY: Alpha Books, 2008.

Ham, Becky. *The Periodic Table* (Essential Chemistry). New York, NY: Chelsea House Publishers, 2008.

Johnson, Rebecca. *Atomic Structure* (Great Ideas of Science). Rev. ed. Minneapolis, MN: Twenty-First Century Books, 2008.

Jones, Maitland Jr., and Steven A. Fleming. *Organic Chemistry*. 4th ed. New York, NY: W. W. Norton & Company, 2009.

Joy, Charles A. "Biographical Sketch of Frederick (sic) Wöhler." *Popular Science Monthly*, 1880. Today in Science History Web site. Retrieved June 27, 2013 (http://todayinsci.com/W/Wohler_Friedrich/Wohler Friedrich-BioSketch(1880).htm).

Keen, Robin. "Friedrich Wöhler and His Lifelong Interest in the Platinum Metals." *Platinum Metals Review*, 1985, Vol. 29, No. 2, pp. 81–85. Retrieved June 27, 2013 (http://www

.platinummetalsreview.com/pdf/pmr-v29-i2-081-085.pdf).

Manning, Phillip. *Chemical Bonds* (Essential Chemistry). New York, NY: Chelsea House Publishers, 2009.

Miller, Ron. *The Elements. What You Really Want to Know*. Minneapolis, MN: Twenty-First Century Books, 2006.

Rogers, Kara, ed. *The Chemical Reactions of Life: From Metabolism to Photosynthesis* (Biochemistry, Cells, and Life). New York, NY: Britannica Educational Publishing and Rosen Educational Services, 2011.

Rogers, Kara, ed. *The Components of Life: From Nucleic Acids to Carbohydrates* (Biochemistry, Cells, and Life). New York, NY: Britannica Educational Publishing and Rosen Educational Services, 2011.

Roza, Greg. *The Halogen Elements* (Understanding the Elements of the Periodic Table). New York, NY: Rosen Publishing, 2010.

Saunders, Nigel. *Atoms and Molecules* (Exploring Physical Science). New York, NY: Rosen Publishing, 2008.

Saunders, Nigel. *Chemical Reactions* (Exploring Physical Science). New York, NY: Rosen Publishing, 2008.

Saunders, Nigel. *Fluorine and the Halogens* (The Periodic Table). Chicago, IL: Heinemann, 2005.

Thorpe, T. E. "Freidrich Wöhler." Edited version of essay original published in *Essays in Historical Chemistry*. New York, NY, and London, England: Macmillan Ltd., 1902. Chemistry in Action Web site. Retrieved June 13, 2013 (http://www3.ul.ie/~childsp/CinA/Issue60/TOC22_Wohler.htm).

West, Krista. *Carbon Chemistry* (Essential Chemistry). New York, NY: Chelsea House Publishers, 2008.

Williams, Blair, and Michael P. Clough, Matthew Stanley, and James T. Colbert. "A Distinctly Human Quest: The Demise of Vitalism and the Search for Life's Origin." The Story Behind the Science Web site. Retrieved June 27, 2013 (http://www.storybehindthescience.org/pdf/vitalism.pdf).

Wolny, Philip. *Chemical Reactions* (Science Made Simple). New York, NY: Rosen Publishing, 2011.

PHOTO CREDITS

Cover, p. 3, interior pages (borders) © iStockphoto.com/Oxford; pp. 6, 11, 14, 15 (top, bottom), 17, 21, 25 iStockphoto/Thinkstock; p. 7 (top) Leemage/Universal Images Group/Getty Images; p. 7 (bottom) SeDmi/Shutterstock.com; p. 9 (top) JanBussan/Shutterstock.com; p. 9 (bottom) Le Do/Shutterstock.com; p. 10 RTimages/Shutterstock.com; p. 12 © iStockphoto.com/Mordolff; p. 16 Igor Plotnikov/Shutterstock.com; p. 18 sgame/Shutterstock.com; p. 19 Jiang Hongyan/Shutterstock.com; p. 20 Microstock Man/Shutterstock.com; p. 23 Stocktrek Images/Fahad Sulehria/Getty Images; p. 24 think4photop/Shutterstock.com; p. 25 (top inset) Garsya/Shutterstock.com; p. 27 Bigone/Shutterstock.com; p. 28 (top) Paul Rapson/Science Source; p. 28 (bottom) Jelle vd Wolf/Shutterstock.com; p. 29 Eye of Science/Science Source; p. 30 (top) Paladin12/Shutterstock.com; p. 30 (bottom) Scott J. Carson/Shutterstock.com; pp. 31 (top), 42 Monty Rakusen/Cultura/Getty Images; p. 31 (bottom) AFP/Getty Images; p. 32 Marie C Fields/Shutterstock.com; p. 33 Oleg Golovnev/Shutterstock.com; p. 35 hxdbzxy/Shutterstock.com; p. 36 Stocktrek Images/Getty Images; pp. 37, 78 Hulton Archive/Getty Images; p. 38 (top) Federico Rostagno/Shutterstock.com; p. 38 (bottom) Laguna Design/Science Photo Library/Getty Images; p. 39 (top) Jeffrey Coolidge/Iconica/Getty Images; p. 39 (bottom) Geoff Tompkinson/Science Source; p. 40 Ramon L. Farinos/Shutterstock.com; p. 43 Lebazele/E+/Getty Images; p. 44 (top) Martyn F. Chillmaid/Science Source; p. 44 (bottom) Print Department, Boston Public Library; p. 45 (top) Science & Society Picture Library/Getty Images; p. 45 (bottom) Viktar Malyshchyts/Shutterstock.com; p. 46 (bottom left) James H Robinson/Photo Researchers/Getty Images; p. 46 (bottom right) Dr. Kenneth Greer/Visuals Unlimited/Getty Images; p. 47 Marko Poplasen/Shutterstock.com; p. 48 laranik/Shutterstock.com; p. 49 Svitlana-ua/Shutterstock.com; p. 50 (top) Mihai Simonia/Shutterstock.com; p. 50 (bottom) Kallista Images/Getty Images; p. 51 Cary Wolinsky/Aurora/Getty Images; p. 53 Big Cheese Photo/Thinkstock; p. 54 (top) M K Ranjitsinh/Photo Researchers/Getty Images; p. 54 (bottom) Romanchuck Dimitry/Shutterstock.com; p. 55 (top) stavklem/Shutterstock.com; p. 55 (bottom) Alta Oosthuizen/Shutterstock.com; p. 56 (top) Heiko Kiera/Shutterstock.com; p. 56 (bottom) Oleksiy Avtomonov/Shutterstock.com; p. 59 (top) Roger Charity/StockImage/Getty Images; p. 59 (middle) Science and Society/SuperStock; p. 59 (bottom) Derek Berwin/Photographer's Choice/Getty Images; p. 60 holbox/Shutterstock.com; p. 61 Ljupco Smokovski/Shutterstock.com; p. 62 (top) Sports Studio Photos/Getty Images; p. 62 (bottom), p. 69 stoonn/Shutterstock.com; p. 64 (top) Charles D Winters/Photo Researchers/Getty Images; p.64 (bottom) Digital Media Pro/Shutterstock.com; p. 66 (top) KKulikov/Shutterstock.com; p. 66 (bottom) Krasowit/Shutterstock.com; p. 67 Huguette Roe/Shutterstock.com; p. 68 Steve Cukrov/Shutterstock.com; p. 71 Sergio Schnitzler/Shutterstock.com; p. 72 Guy Vanderelst/Photographer's Choice RF/Getty Images; p. 74 llizia/Shutterstock.com; p. 75 Universal Images Group/Getty Images; p. 77

Igor Golovniov/Shutterstock.com; p. 79 De Agostini/Getty Images; p. 80 Bruno Kussler Marques/Wikimedia Commons/File: Goettingen Woehler Statue.jpg/CC BY 2.0; pp. 82–83 Kris Everson; top of pp. 6, 10, 20, 32, 40, 48, 60, 72 © iStockphoto.com/aleksandarvelasevic; all other photos and illustrations © Brown Bear Books Ltd.